John Dawson Ross

Round Burns' Grave

The paeans and dirges of many bards

John Dawson Ross

Round Burns' Grave
The paeans and dirges of many bards

ISBN/EAN: 9783337328849

Printed in Europe, USA, Canada, Australia, Japan

Cover: Foto ©berggeist007 / pixelio.de

More available books at **www.hansebooks.com**

ROUND BURNS' GRAVE:

THE

Paeans and Dirges of many Bards,

GATHERED TOGETHER BY

JOHN D. ROSS,

EDITOR OF "CELEBRATED SONGS OF SCOTLAND," AND AUTHOR OF
"SCOTTISH POETS IN AMERICA."

NEW AND ENLARGED EDITION.

PAISLEY : ALEXANDER GARDNER,
Publisher to Her Majesty the Queen.

1892.

Round Burns' Grave, a poet band—
Singers, not of his native land
Alone—but bards of every clime,
Salute the Poet of all time.

And each his loving tribute lays—
A wreath of cypress twin'd with bays;
As they approach the tomb by turns—
That holds the sacred dust of Burns—

Feeble they feel their tongues to sing
The Praises of their Poet King;
But in each heart a quenchless flame
Leaps up to greet the Poet's name!

Perchance his spirit hovering near
May stoop these lays of love to hear,
And breathe once more its magic spell
O'er brother bards who love him well.
—James D. Crichton.

DEDICATED TO

Thomas C. Latto, F.B.H.S.,

(Author of " The Kiss ahint the Door," " When we were at the Schule," and various other well-known Scottish Songs and Poems),

IN ACKNOWLEDGEMENT OF MANY FAVOURS
RECEIVED AT HIS HANDS.

J. D. R.

CONTENTS.

	PAGE
Burns. By Fitz-Greene Halleck,	9
Ode on the Centenary of Burns. By Isabella Craig Knox,	16
On the Death of Burns. By William Roscoe,	20
Robert Burns. By Henry W. Longfellow,	25
Ode to the Memory of Burns. By Thos. Campbell,	27
The Gift of Burns. By Robert Buchanan,	31
For the Burns Centennial Celebration, January 25, 1859. By Oliver Wendell Holmes,	34
Burns. By John G. Whittier,	37
An Incident in a Railroad Car. By James Russell Lowell,	42
Thoughts. By William Wordsworth,	46
To the Sons of Burns. By William Wordsworth,	49
Mossgiel. By Wm. Wordsworth,	51
Burns at Mossgiel. By Charles Kent,	52
Robert Burns. By James Montgomery,	56
For the Centenary of Robert Burns. By Robert Leighton,	58
Address to the Shade of Burns. By Capt. Charles Gray,	61
Robert Burns. By Dr. John M. Harper,	65
Rantin' Robin, Rhymin' Robin. By David Vedder,	69
Ellisland. By Prof. John Stuart Blackie,	71
Burns' Birthday. By Thos. Buchanan Read,	73
Lines written on the Anniversary of Burns' Birthday. By Hugh Ainslie,	77
Burns. By John Imlah,	79
Robin's Awa! By James Hogg,	82
Ode written for the Celebration of Robert Burns' Birthday, Paisley. By Robert Tannahill,	84
Robert Burns. By Thomas Fraser,	88
The Bard of Song. By Robert Gilfillan,	96
Ode for the Anniversary of Robert Burns. By Wm. Glen,	98

CONTENTS.

	PAGE
Verses written on visiting the House in which Robert Burns was born. By Richard Gall,	101
Coila's Bard. By James Stirrat,	103
Elegy to the Memory of Robert Burns. By Alexander Balfour,	106
What is Success? Or the Philosopher and the Ploughman. By Thos. C. Latto,	114
Burns. By William Murray,	120
The Birthplace of Robert Burns. By Robert G. Ingersoll,	121
A Poet King. By John Macfarlane,	122
Rantin' Robin. A Song for Burns' Anniversary. By A. H. Wingfield,	124
To the Memory of Burns. By Francis Bennoch,	126
Address to Burns. By James D. Crichton,	128
To the Memory of Robert Burns. By Edward Rushton,	133
Robert Burns. By Evan MacColl,	137
On the Death of Burns. By Mrs. Grant of Laggan,	139
Stanzas written on a copy of the Engraving of Robert Burns. By Thomas Atkinson,	142
Written for Burns' Anniversary. By Robert Allan,	144
Thoughts on visiting the Grave of Burns. By Alexander Maclaggan,	146

	PAGE
Song for the Anniversary of the Birthday of Robert Burns. By Andrew Park,	149
Robert Burns. By Joseph Cunningham,	151
Birthplace of Robert Burns. By Thomas W. Parsons,	152
To the Memory of Robert Burns. By James Macfarlan,	154
Ye may talk o' your Learning. By Andrew Mercer,	157
The night you quoted Burns to me. By James Newton Matthews,	159
Birth of Burns. By Thos. Miller,	161
An Evening with Burns. By Agnes Maule Machar,	167
Kossuth at the Grave of Burns. By Alexander G. Murdoch,	170
The Burns Monument, Kilmarnock. By Alexander G. Murdoch,	174
Stanzas for the Burns Festival at Ayr, 1844. By David Macbeth Moir,	178
Wild Flowers from Alloway and Doon. By Alexander Anderson,	182
Robert Burns. By Alexander Anderson,	185
Burns' Vision of the Future. By Miles Macphail,	192
The Poet's Jubilee. By Thomas C. Latto,	197
Robert Burns. By W. Stewart Ross,	205
To Burns. By J. M'Intosh,	209
The Soul of Burns in Song. By Hon. W. C. Sturoc,	211
A Ploughman's Dream. By Robert Hogg,	213

CONTENTS.

	PAGE
Kossuth at the Grave of Burns. By W. Stewart Ross,	221
On Visiting the Tomb of Robert Burns. By W. Hogg,	225
A Burns Anniversary Song. By Colin Rae Brown,	227
Lines Written in Burns' Cottage. By Rev. R. S. Bowie,	229
At the Grave of Robert Burns. By Charles Mackay, LL.D.,	230
To the Memory of Robert Burns. By William Reid,	232
Robert Burns. By John Macfarlane,	238
To Robert Burns. By John Patterson,	239
The Grave of Burns. By Robert Nicoll,	242
Ode written for the Celebration of Robert Burns' Birthday, Paisley. By Robert Tannahill,	244
Burns Remembered. By Rev. Arthur J. Lockhart,	247
Robin Burns. By Robert Ford,	250
Burns. By David Wingate,	253
On Anniversary of Birthday of Robert Burns. By Sarah Parker Douglas,	256
Burns on his Death-bed. By W. M'Dowall,	259
Robert Burns. By Malcolm Taylor, Jr.,	262
To Burns. By David Vedder,	266
For the Anniversary of Burns. By David Vedder,	268
Freedom's Bard. By John Kelso Kelly,	270
The Harp of Burns. By Alexander Maclagan,	273
Robert Burns: a Centenary Ode. By Rev. William Buchanan, B.A.,	275
The Auld Brig's Welcome. By Wallace Bruce,	279
Dinna Forget. By Hunter MacCulloch,	284
Lines on Robert Burns. By Rev. John Burtt,	287
Ode for the Burns Anniversary. By William Thomson,	288
Coila. By Francis Bennoch,	291
On the Death of Burns. By Richard Gall,	294
On Burns Anniversary. By Hew Ainslie,	296
Lines written for Burns' Anniversary. By John Mitchell,	298
Robert Burns: a Centenary Song. By Gerald Massey,	301
Robert Burns. By William Freeland,	315

ROUND BURNS' GRAVE.

Burns.

Fitz-Greene Halleck.

To a Rose, brought from near Alloway Kirk, in Ayrshire, in the Autumn of 1822.

Wild Rose of Alloway! my thanks:
   Thou 'mindst me of that autumn noon
When first we met upon " the banks
   And braes o' bonny Doon."

Like thine, beneath the thorn-tree's bough,
   My sunny hour was glad and brief;
We've crossed the winter sea, and thou
   Art withered—flower and leaf.

And will not thy death-doom be mine—
   The doom of all things wrought of clay—
And withered my life's leaf like thine,
   Wild Rose of Alloway?

Not so his memory, for whose sake
  My bosom bore thee far and long;
His—who a humbler flower could make
  Immortal as his song.

The memory of Burns!—a name
  That calls, when brimmed her festal cup,
A nation's glory and her shame,
  In silent sadness up.

A nation's glory!—be the rest
  Forgot—she's canonized his mind;
And it is joy to speak the best
  We may of human kind.

I've stood beside the cottage bed
  Where the Bard-peasant first drew breath;
A straw-thatched roof above his head,
  A straw-wrought couch beneath.

And I have stood beside the pile,
  His monument—that tells to Heaven
The homage of earth's proudest isle
  To that Bard-peasant given!

Bid thy thoughts hover o'er that spot,
  Boy-minstrel, in thy dreaming hour;
And know, however low his lot,
  A Poet's pride and power.

BURNS.

The pride that lifted Burns from earth,—
The power that gave a child of song
Ascendency o'er rank and birth,
The rich, the brave, the strong;

And if despondency weigh down
Thy spirit's fluttering pinions then,
Despair—thy name is written on
The roll of common men.

There have been loftier themes than his,
And longer scrolls, and louder lyres,
And lays lit up with Poesy's
Purer and holier fires:

Yet read the names that know not death;
Few nobler ones than Burns are there;
And few have won a greener wreath
Than that which binds his hair.

His is that language of the heart,
In which the answering heart would speak,—
Thought, word, that bids the warm tear start,
Or the smile light the cheek;

And his that music, to whose tone
The common pulse of man keeps time,
In cot or castle's mirth or moan,
In cold or sunny clime.

And who hath heard his song, nor knelt
   Before its spell with willing knee,
And listened, and believed, and felt
   The Poet's mastery.

O'er the Mind's sea, in calm or storm,
   O'er the Heart's sunshine and its showers,
O'er Passion's moments, bright and warm,
   O'er Reason's dark, cold hours;

On fields where brave men "die or do,"
   In halls where rings the banquet's mirth,
Where mourners weep, where lovers woo,
   From throne to cottage hearth?

What sweet tears dim the eyes unshed
   What wild vows falter on the tongue,
When "Scots wha hae wi' Wallace bled,"
   Or "Auld Lang Syne" is sung!

Pure hopes, that lift the soul above,
   Come with his Cottar's hymn of praise,
And dreams of youth, and truth, and love,
   With "Logan's" banks and braes.

And when he breathes his master-lay
   Of Alloway's witch-haunted wall,
All passions in our frames of clay
   Come thronging at his call.

Imagination's world of air,
    And our own world, its gloom and glee,
Wit, pathos, poetry, are there,
    And death's sublimity.

And Burns—though brief the race he ran,
    Though rough and dark the path he trod—
Lived—died—in form and soul a Man,—
    The image of his God.

Through care, and pain, and want and woe,
    With wounds that only death could heal;
Tortures—the poor alone can know,
    The proud alone can feel;

He kept his honesty and truth,
    His independent tongue and pen,
And moved, in manhood as in youth,
    Pride of his fellow men.

Strong sense, deep feeling, passions strong,
    A hate of tyrant and of knave;
A love of right, a scorn of wrong,
    Of coward and of slave;

A kind, true heart, a spirit high,
    That could not fear and would not bow,
Were written in his manly eye
    And on his manly brow.

Praise to the bard! his words are driven,
    Like flower-seeds by the far wind sown,
Where'er, beneath the sky of heaven,
    The birds of fame have flown.

Praise to the man! a nation stood
    Beside his coffin with wet eyes,—
Her brave, her beautiful, her good,
    As when a loved one dies.

And still, as on his funeral day,
    Men stand his cold earth-couch around,
With the mute homage that we pay
    To consecrated ground.

And consecrated ground it is,—
    The last, the hallowed home of one
Who lives upon all memories,
    Though with the buried gone.

Such graves as his are pilgrim-shrines,—
    Shrines to no code or creed confined—
The Delphian vales, the Palestines,
    The Meccas of the mind.

Sages, with wisdom's garland wreathed,
    Crowned kings, and mitred priests of power,
And warriors with their bright swords sheathed,
    The mightiest of the hour;

And lowlier names, whose humble home
  Is lit by Fortune's dimmer star,
Are there—o'er wave and mountain come,
  From countries near and far;

Pilgrims whose wandering feet have pressed
  The Switzer's snow, the Arab's sand,
Or trod the piled leaves of the West,—
  My own green forest-land.

All ask the cottage of his birth,
  Gaze on the scenes he loved and sung,
And gather feelings not of earth
  His fields and streams among.

They linger by the Doon's low trees,
  And pastoral Nith, and wooded Ayr,
And round thy sepulchres, Dumfries!
  The poet's tomb is there.

But what to them the sculptor's art,
  His funeral columns, wreaths, and urns?
Were they not graven on the heart—
  The name of Robert Burns!

Ode on the Centenary of Burns.

Isabella Craig Knox.

WE hail this morn,
A century's noblest birth ;
A poet peasant-born,
Who more of Fame's immortal dower
  Unto his country brings
  Than all her kings !

  As lamps high set
Upon some earthly eminence ;
And to the gazer brighter thence
Than the sphere lights they flout—
  Dwindle in distance and die out,
  While no star waneth yet ;
So through the past's far-reaching night
  Only the star-souls keep their light.

  A gentle boy,
With moods of sadness and of mirth,
  Quick tears and sudden joy,
Grew up beside the peasant's hearth.
  His father's toil he shares ;
  But half his mother's cares
  From his dark, searching eyes,
Too swift to sympathise,
  Hid in her heart she bears.

  At early morn
His father calls him to the field ;

Through the stiff soil that clogs his feet,
  Chill rain, and harvest heat
He plods all day; returns at eve outworn,
  To the rude fare a peasant's lot doth yield—
To what else was he born?

  The God-made king
  Of every living thing;
(For his great heart in love could hold them all);
The dumb eyes meeting his by hearth and stall—
  Gifted to understand!—
  Knew it and sought his hand;
And the most timorous creature had not fled
  Could she his heart have read,
Which fain all feeble things had blessed and sheltered.

  To Nature's feast,
Who knew her noblest guest
And entertained him best,
Kingly he came. Her chambers of the east
  She draped with crimson and with gold,
    And poured her pure joy wines
      For him the poet-souled;
      For him her anthem rolled
From the storm-wind among the winter pines,
    Down to the slenderest note
Of a love-warble from the linnet's throat.

  But when begins
The array for battle, and the trumpet blows,
A king must leave the feast and lead the fight;
  And with its mortal foes,
Grim gathering hosts of sorrows and of sins,
  Each human soul must close;

And Fame her trumpet blew
Before him, wrapped him in her purple state,
And made him mark for all the shafts of Fate
   That henceforth round him flew.

Though he may yield,
Hard-pressed, and wounded fall
   Forsaken on the field ;
   His regal vestments soiled ;
   His crown of half its jewels spoiled ;
     He is a king for all.
   Had he but stood aloof !
Had he arrayed himself in armour proof
   ' Against temptation's darts !
So yearn the good—so those the world calls wise,
   With vain, presumptuous hearts,
     Triumphant moralise.

Of martyr-woe
A sacred shadow on his memory rests—
   Tears have not ceased to flow—
Indignant grief yet stirs impetuous breasts,
   To think—above that noble soul brought low,
That wise and soaring spirit fooled, enslaved—
     Thus, thus he had been saved !

   It might not be !
That heart of harmony
   Had been too rudely rent ;
Its silver chords, which any hand could wound,
   By no hand could be tuned,
Save by the Maker of the instrument,
   Its every string who knew,
And from profaning touch His heavenly gift withdrew.

Regretful love
His country fain would prove,
By grateful honours lavished on his grave ;
Would fain redeem her blame
That he so little at her hands can claim,
Who unrewarded gave
To her his life-bought gift of song and fame.

The land he trod
Hath now become a place of pilgrimage ;
Where dearer are the daisies of the sod
That could his song engage.
The hoary hawthorn, wreathed
Above the bank on which his limbs he flung
While some sweet plaint he breathed ;
The streams he wandered near ;
The maidens whom he loved ; the songs he sung—
All—all are dear !

The arch blue eyes—
Arch but for love's disguise—
Of Scotland's daughters, soften at his strain ;
Her hardy sons, sent forth across the main
To drive the ploughshare through earth's virgin soils,
Lighten with it their toils ;
And sister-lands have learn'd to love the tongue
In which such songs are sung.

For doth not song
To the whole world belong ?
Is it not given wherever tears can fall,
Wherever hearts can melt, or blushes glow,
Or mirth and sadness mingle as they flow,
A heritage to all ?

On the Death of Burns.

WILLIAM ROSCOE.

REAR high thy bleak majestic hills,
   Thy sheltered valleys proudly spread,
And, SCOTIA, pour thy thousand rills,
   And wave thy heaths with blossoms red.
But ah! what poet now shall tread
   Thy airy heights, thy woodland reign,
Since he, the sweetest bard, is dead,
   That ever breath'd the soothing strain!

As green thy towering pines may grow,
   As clear thy streams may speed along,
As bright thy summer suns may glow,
   As gaily charm thy feathery throng;
But now, unheeded is the song,
   And dull and lifeless all around,
For his wild harp lies all unstrung,
   And cold the hand that waked its sound.

What tho' thy vigorous offspring rise
   In arts, in arms, thy sons excel;
Tho' beauty in thy daughters' eyes,
   And health in every feature dwell;

ON THE DEATH OF BURNS.

Yet who shall now their praises tell,
   In strains impassion'd, fond and free,
Since he no more the song shall swell
   To love, and liberty, and thee.

With step-dame eye and frown severe
   His hapless youth why didst thou view?
For all thy joys to him were dear,
   And all his vows to thee were due;
Nor greater bliss his bosom knew,
   In opening youth's delightful prime,
Than when thy favouring ear he drew
   To listen to his chanted rhyme.

Thy lonely wastes and frowning skies
   To him where all with rapture fraught;
He heard with joy the tempest rise
   That waked him to sublimer thought;
And oft thy winding dells he sought,
   Where wild flow'rs pour'd their rich perfume,
And with sincere devotion brought
   To thee the summer's earliest bloom.

But ah! no fond maternal smile,
   His unprotected youth enjoy'd,
His limbs inur'd to early toil,
   His days with early hardships tried:
And more to mark the gloomy void,
   And bid him feel his misery,
Before his infant eyes would glide
   Day dreams of immortality.

Yet, not by cold neglect depress'd,
  With sinewy arm he turn'd the soil,
Sunk with the evening sun to rest,
  And met at morn his earliest smile.
Waked by his rustic pipe, meanwhile
  The powers of fancy came along,
And sooth'd his lengthen'd hours of toil,
  With native wit and sprightly song.

—Ah! days of bliss, too swiftly fled,
  When vigorous health from labour springs,
And bland contentment smoothes the bed,
  And sleep his ready opiate brings;
And hovering round on airy wings
  Float the light forms of young desire,
That of unutterable things
  The soft and shadowy hope inspire.

Now spells of mightier power prepare,
  Bid brighter phantoms round him dance;
Let Flattery spread her viewless snare,
  And Fame attract his vagrant glance;
Let sprightly Pleasure too advance,
  Unveil'd her eyes, unclasp'd her zone,
Till, lost in love's delirious trance,
  He scorns the joys his youth has known.

Let Friendship pour her brightest blaze,
  Expanding all the bloom of soul;
And Mirth concentre all her rays,
  And point them from the sparkling bowl;

And let the careless moments roll
   In social pleasure unconfined,
And confidence that spurns control
   Unlock the inmost springs of mind :

And lead his steps those bowers among,
   Where elegance with splendour vies,
Or Science bids her favour'd throng,
   To more refined sensations rise :
Beyond the peasant's humbler joys,
   And freed from each laborious strife,
There let him learn the bliss to prize
   That waits the sons of polish'd life.

Then whilst his throbbing veins beat high
   With every impulse of delight,
Dash from his lips the cup of joy,
   And shroud the scene in shades of night ;
And let Despair, with wizard light,
   Disclose the yawning gulf below,
And pour incessant on his sight
   Her spectred ills and shapes of woe :

And show beneath a cheerless shed,
   With sorrowing heart and streaming eyes,
In silent grief where droops her head,
   The partner of his early joys ;
And let his infants' tender cries
   His fond parental succour claim,
And bid him hear in agonies
   A husband's and a father's name.

ON THE DEATH OF BURNS.

'Tis done, the powerful charm succeeds ;
   His high reluctant spirit bends ;
In bitterness of soul he bleeds,
   Nor longer with his fate contends.
An idiot laugh the welkin rends
   As genius thus degraded lies ;
Till pitying Heaven the veil extends
   That shrouds the Poet's ardent eyes.

—Rear high thy bleak majestic hills,
   Thy sheltered valleys proudly spread,
And SCOTIA, pour thy thousand rills,
   And wave thy heath with blossoms red :
But never more shall poet tread
   Thy airy height, thy woodland reign,
Since he, the sweetest bard, is dead,
   That ever breath'd the soothing strain.

Robert Burns.

HENRY W. LONGFELLOW.

I SEE amid the fields of Ayr
A ploughman, who, in foul and fair,
    Sings at his task
So clear, we know not if it is
The laverock's song we hear, or his,
    Nor care to ask.

For him the ploughing of those fields
A more ethereal harvest yields
    Than sheaves of grain;
Songs flush with purple bloom the rye,
The plover's call, the curlew's cry
    Sing in his brain.

Touched by his hand, the wayside weed
Becomes a flower; the lowliest reed
    Beside the stream
Is clothed in beauty; gorse and grass
And heather, where his footsteps pass,
    The brighter seem.

He sings of love, whose flame illumes
The darkness of lone cottage rooms;
    He feels the force,
The treacherous undertow and stress
Of wayward passions, and no less
    The keen remorse.

At moments, wrestling with his fate,
His voice is harsh, but not with hate;
    The brushwood, hung
Above the tavern door, lets fall
Its bitter leaf, its drops of gall
    Upon his tongue.

But still the music of his song
Rises o'er all elate and strong;
    Its master-chords
Are Manhood, Freedom, Brotherhood;
Its discords but an interlude
    Between the words.

And then to die so young and leave
Unfinished what he might achieve
    Yet better sure
Is this, than wandering up and down
An old man in a country town,
    Infirm and poor.

For now he haunts his native land
As an immortal youth; his hand
    Guides every plough;
He sits beside each ingle-nook,
His voice is in each rushing brook,
    Each rustling bough.

His presence haunts this room to-night,
A form of mingled mist and light
    From that far coast.
Welcome beneath this roof of mine!
Welcome! this vacant chair is thine,
    Dear guest and ghost!

Ode to the Memory of Burns.

THOMAS CAMPBELL.

Soul of the Poet ! whereso'er
Reclaim'd from earth, thy genius plume
Her wings of immortality !
Suspend thy harp in happier sphere,
And with thy influence illume
The gladness of our jubilee.

And fly like fiends from secret spell,
Discord and strife, at BURNS' name,
Exorcised by his memory?
For he was chief of bards that swell
The heart with songs of social flame,
And high delicious revelry.

And Love's own strain to him was given,
To warble all its ecstasies
With Pythian words unsought, unwill'd,—
Love, the surviving gift of Heaven,
The choicest sweet of Paradise,
In life's else bitter cup distill'd.

Who that has melted o'er his lay
To Mary's soul, in Heaven above,
But pictured sees, in fancy strong,
The landscape and the livelong day
That smiled upon their mutual love?
Who that has felt forgets the song?

Nor skill'd one flame alone to fan;
His country's high soul'd peasantry
What patriot pride he taught!—how much
To weigh the inborn worth of man!
And rustic life and poverty
Grow beautiful beneath his touch.

Him, in his clay-built cot, the Muse
Entranced, and showed him all the forms
Of fairy-light and wizard gloom,
(That only gifted poet views)
The Genii of the floods and storms,
And martial shades from Glory's tomb.

On Bannock-field, what thoughts arouse
The swain whom BURNS's song inspires;
Beat not his Caledonian veins,
As o'er the heroic turf he ploughs,
With all the spirit of his sires,
And all their scorn of death and chains?

And see the Scottish exile, tann'd
By many a far and foreign clime,

ODE TO THE MEMORY OF BURNS.

Bend o'er his home-born verse, and weep
In memory of his native land,
With love that scorns the lapse of time,
And ties that stretch beyond the deep.

Encamp'd by Indian rivers wild,
The soldier resting on his arms,
In BURNS's carol sweet recals
The scenes that bless'd him when a child,
And glows and gladdens at the charms
Of Scotia's woods and waterfalls.

O deem not, 'midst the worldly strife,
An idle art the Poet brings :
Let high Philosophy control,
And sages calm, the stream of life,
'Tis he refines its fountain-springs,—
The nobler passions of the soul.

It is the Muse that consecrates
The native banner of the brave,
Unfurling, at the trumpet's breath,
Rose, thistle, harp; 'tis she elates
To sweep the field or ride the wave,—
A sunburst in the storm of death.

And thou, young hero, when thy pall
Is cross'd with mournful sword and plume,
When public grief begins to fade,
And only tears of kindred fall,
Who but the bard shall dress thy tomb,
And greet with fame thy gallant shade?

ODE TO THE MEMORY OF BURNS.

Such was the soldier—BURNS, forgive
That sorrows of mine own intrude
In strains to thy great memory due.
In verse like thine, oh ! could he live,
The friend I mourn'd—the brave—the good—
Edward that died at Waterloo !*

Farewell, high chief of Scottish song !
That couldst alternately impart
Wisdom and rapture in thy page,
And brand each vice with satire strong ;
Whose lines are mottoes of the heart—
Whose truths electrify the sage.

Farewell ! and ne'er may Envy dare
To wring one baleful poison drop
From the crush'd laurels of thy bust :
But while the lark sings sweet in air,
Still may the grateful pilgrim stop,
To bless the spot that holds thy dust.

* Major Edward Hodge, of the 7th Hussars, who fell at the head of his squadron in the attack of the Polish Lancers.

The Gift of Burns.

ROBERT BUCHANAN.

Addressed to the Boston Caledonian Club on the one hundred and twenty-sixth anniversary of the Birth of the National Poet.

I.

THAT speech the English Pilgrims spoke
    Fills the great plains afar,
And branches of the British vale
    Wave 'neath the Western star ;
" Be free ! " men cried, in Shakespeare's tongue,
    When striking for the slave—
Thus Hampden's cry for Freedom rung
    As far as Lincoln's grave !

II.

But when new vales of England rise,
    The thistle freelier blows ;
Across the seas 'neath alien skies
    Another Scotland grows ;
Here Independence, mountain maid,
    Reaps her full birthright now,
And BURNS'S shade, in trews and plaid,
    Still whistles at the plough.

III.

Scots, gather'd now in phalanx bright,
    Here in this distant land,
To greet you, this immortal night,
    I reach the loving hand ;
My soul is with you, one and all,
    Who pledge our poet's fame,
And echoing your toast, I call
    A blessing on his name !

IV.

The heritage he left behind
    Has spread from sea to sea—
The liberal heart, the fearless mind,
    The undaunted soul and free ;
The radiant humour that redeem'd
    A world of commonplace ;
The wit that like a sword-flash gleam'd
    In Fashion's painted face ;

V.

The brotherhood where smiles and tears,
    Too deep for thought to scan,
Have made of all us mountaineers
    One world-compelling clan !
Hand join with hand. Soul links with soul
    Where'er we sit and sing,
Flashing from utmost pole to pole,
    Love's bright electric ring !

VI.

The songs he sang were sown as seeds
    Sown in the furrow'd earth—
They ripen into dauntless deeds,
    And flowers of gentle mirth ;

They brighten every path we tread,
  They conquer time and place;
While blue skies, opening overhead,
  Reveals—the singer's face!

VII.

He struggled, agonized, and fell
  As all who live have thriven,
But with his wit he conquer'd Hell,
  And with his love show'd Heaven!
He was the noblest of us all,
  Yet of us all a part,
For every Scot, howe'er so small,
  Is high as BURNS'S heart!

VIII.

Immortal is the night, indeed,
  When he this life began—
The open-handed, stubborn-knee'd,
  Type of the mountain clan!
The shape erect that never knelt
  To Kings of earth or air,
But at a maiden's touch would melt,
  And tremble into prayer!

IX.

His soul pursues us where we roam,
  Beyond the furthest waves,
He sheds the light of love and home
  Upon our loneliest graves!
Poor is the slave that honours not
  The flag *he* first unfurl'd—
Our singer, who has made the Scot
  The freeman of the world!

For the Burns Centennial Celebration, January 25, 1859.

OLIVER WENDELL HOLMES.

His birthday.—Nay, we need not speak
    The name each heart is beating,—
Each glistening eye and flushing cheek
    In light and flame repeating!

We come in our tumultuous tide,—
    One surge of wild emotion,—
As crowding through the Frith of Clyde
    Rolls in the Western Ocean;

As when yon cloudless, quartered moon
    Hangs o'er each storied river,
The swelling breasts of Ayr and Doon
    With sea-green wavelets quiver.

The century shrivels like a scroll,—
    The past becomes the present,—
And face to face, and soul to soul,
    We greet the monarch-peasant.

While Shenstone strained in feeble flights
    With Corydon and Phyllis,—
While Wolfe was climbing Abraham's heights
    To snatch the Bourbon lilies,—

Who heard the wailing infant's cry,
    The babe beneath the sheiling,
Whose song to-night in every sky
    Will shake earth's starry ceiling,—

Whose passion-breathing voice ascends
    And floats like incense o'er us,
Whose ringing lay of friendship blends
    With labour's anvil chorus?

We love him, not for sweetest song,
    Though never tone so tender;
We love him, even in his wrong,—
    His wasteful self-surrender.

We praise him, not for gifts divine,—
    His Muse was born of woman,—
His manhood breathes in every line,—
    Was ever heart more human?

We love him, praise him, just for this:
    In every form and feature,
Through wealth and want, through woe and bliss,
    He saw his fellow-creature?

No soul could sink beneath his love,—
　　Not even angel blasted ;
No mortal power could soar above
　　The pride that all outlasted !

Ay ! Heaven had set one living man
　　Beyond the pedant's tether,—
His virtues, frailties, HE may scan,
　　Who weighs them altogether !

I fling my pebble on the cairn
　　Of him, though dead, undying ;
Sweet Nature's nursling, bonniest bairn
　　Beneath her daisies lying.

The waning suns, the wasting globe,
　　Shall spare the minstrel's story,—
The centuries weave his purple robe,
　　The mountain-mist of glory !

Burns.

On receiving a sprig of heather in blossom.

JOHN G. WHITTIER.

No more these simple flowers belong
   To Scottish maid and lover;
Sown in the common soil of song,
   They bloom the wide world over.

In smile and tears, in sun and showers,
   The minstrel and the heather,
The deathless singer and the flowers
   He sang of live together.

Wild heather-bells and ROBERT BURNS!
   The moorland flower and peasant!
How, at their mention, memory turns!
   Her pages old and pleasant!

The gray sky wears again its gold
   And purple of adorning,
And manhood's noonday shadows hold
   The dews of boyhood's morning.

The dews that washed the dust and soil
   From off the wings of pleasure,
The sky, that flecked the ground of toil
   With golden threads of leisure.

I call to mind the summer day,
   The early harvest mowing,
The sky with sun and clouds at play,
   And flowers with breezes blowing.

I hear the blackbird in the corn,
   The locust in the haying;
And, like the fabled hunter's horn,
   Old tunes my heart is playing.

How oft that day, with fond delay,
   I sought the maple's shadow,
And sang with BURNS the hours away,
   Forgetful of the meadow!

Bees hummed, birds twittered, overhead
   I heard the squirrels leaping,
The good dog listened while I read,
   And wagged his tail in keeping.

I watched him while in sportive mood,
   I read "The Twa Dogs'" story,
And half believed he understood
   The poet's allegory.

Sweet days, sweet songs!—The golden hours.
   Grew brighter for that singing,
From brook and bird and meadow flowers
   A dearer welcome bringing.

New light on home-seen nature beamed,
  New glory over Woman ;
And daily life and duty seemed
  No longer poor and common.

I woke to find the simple truth
  Of fact and feeling better
Than all the dreams that held my youth
  A still repining debtor :

That Nature gives her handmaid, Art,
  The theme of sweet discoursing ;
The tender idylls of the heart
  In every tongue rehearsing.

Why dream of lands of gold and pearl,
  Of loving knight and lady,
When farmer boy and barefoot girl
  Were wandering there already ?

I saw through all familiar things
  The romance underlying ;
The joys and griefs that plume the wings
  Of Fancy skyward flying.

I saw the same blithe day return,
  The same sweet fall of even,
That rose on wooded Craigie-burn,
  And sank on crystal Devon.

I matched with Scotland's heathery hills
    The sweet brier and the clover ;
With Ayr and Doon, my native rills,
    Their wood-hymns chanting over.

O'er rank and pomp, as he had seen,
    I saw the Man uprising ;
No longer common or unclean,
    The child of God's baptizing !

With clearer eyes I saw the worth
    Of life among the lowly ;
The Bible at his Cotter's hearth
    Had made my own more holy.

And, if at times an evil strain,
    To lawless love appealing,
Broke in upon the sweet refrain
    Of pure and healthful feeling,

It died upon the eye and ear,
    No inward answer gaining ;
No heart had I to see or hear
    The discord and the staining.

Let those who never erred forget
    His worth, in vain bewailings ;
Sweet Soul of Song !—I own my debt
    Uncancelled by his failings !

Lament who will the ribald line
  Which tells his lapse from duty,
How kissed the maddening lips of wine
  Or wanton ones of beauty;

But think, while falls that shade between
  The erring one and Heaven,
That he who loved like Magdalen,
  Like her may be forgiven.

Not his the song whose thunderous chime
  Eternal echoes render—
The mournful Tuscan's haunted rhyme,
  And Milton's starry splendour!

But who his human heart has laid
  To Nature's bosom nearer?
Who sweetened toil like him, or paid
  To love a tribute dearer?

Through all his tuneful art, how strong
  The human feeling gushes!
The very moonlight of his song
  Is warm with smiles and blushes!

Give lettered pomp to teeth of time,
  So "Bonnie Doon" but tarry;
Blot out the Epic's stately rhyme,
  But spare his Highland Mary!

An Incident in a Railroad Car.

James Russell Lowell.

He spoke of Burns : men rude and rough
   Pressed round to hear the praise of one
Whose heart was made of manly, simple stuff,
   As homespun as their own.

And when he read, they forward leaned,
   Drinking with thirsty hearts and ears,
His brook-like songs whom glory never weaned
   From humble smiles and tears.

Slowly there grew a tender awe,
   Sun-like, o'er faces brown and hard,
As if in him who read they felt and saw
   Some presence of the bard.

It was a sight for sin and wrong
   And slavish tyranny to see,—
A sight to make our faith more pure and strong
   In high humanity.

AN INCIDENT IN A RAILROAD CAR.

I thought these men will carry hence
  Promptings their former life above,
And something of a finer reverence
  For beauty, truth, and love.

God scatters love on every side,
  Freely among his children all,
And always hearts are lying open wide,
  Wherein some grains may fall.

There is no wind but soweth seeds
  Of a more true and open life,
Which burst, unlooked-for, into high-souled deeds,
  With wayside beauty rife.

We find within these souls of ours
  Some wild germs of a higher birth,
Which in the poet's tropic heart bear flowers
  Whose fragrance fills the earth.

Within the hearts of all men lie
  These promises of wider bliss,
Which blossom into hopes that cannot die,
  In sunny hours like this.

All that hath been majestical
  In life or death, since time began,
Is native in the simple heart of all,—
  The angel heart of man.

And thus, among the untaught poor,
   Great deeds and feelings find a home,
That cast in shadow all the golden lore
   Of classic Greece and Rome.

O, mighty brother-soul of man,
   Where'er thou art, in low or high,
Thy skiey arches with exulting span
   O'er—roof infinity !

All thoughts that mould the age begin
   Deep down within the primitive soul,
And from the many slowly upward win
   To one who grasps the whole :

In his wide brain the feeling deep
   That struggled on the many's tongue
Swells to a tide of thought, whose surges leap
   O'er the weak thrones of wrong.

All thought begins in feeling,—wide
   In the great mass its base is hid,
And, narrowing up to thought, stands glorified,
   A moveless pyramid.

Nor is he far astray who deems
   That every hope, which rises and grows broad
In the world's heart, by ordered impulse streams
   From the great heart of God.

AN INCIDENT IN A RAILROAD CAR.

God wills, man hopes: in common souls
   Hope is but vague and undefined,
Till from the poet's tongue the message rolls
   A blessing to his kind.

Never did Poesy appear
   So full of heaven to me, as when
I saw how it would pierce through pride and fear
   To the lives of coarsest men.

It may be glorious to write
   Thoughts that shall glad the two or three
High souls, like those far stars that come in sight
   Once in a century;—

But better far it is to speak
   One simple word, which now and then
Shall waken their free nature in the weak
   And friendless sons of men;

To write some earnest verse or line,
   Which, seeking not the praise of art,
Shall make a clearer faith and manhood shine
   In the untutored heart.

He who doth this, in verse or prose,
   May be forgotten in his day,
But surely shall be crowned at last with those
   Who live and speak for aye.

Thoughts

Suggested the day after seeing the Grave of Burns on the Banks of Nith, near the Poet's residence.

WILLIAM WORDSWORTH.

Too frail to keep the lofty vow
That must have followed when his brow
Was wreathed—"The Vision" tells us how—
    With holly spray,
He faltered, drifted to and fro,
    And passed away.

Well might such thoughts, dear sister, throng
Our minds when lingering, all too long
Over the grave of Burns we hung,
    In social grief—
Indulged as if it were a wrong
    To seek relief.

•

But, leaving each unquiet theme
Where gentlest judgments may misdeem,

And prompt to welcome every gleam
    Of good and fair,
Let us beside this limpid stream
    Breathe hopeful air.

Enough of sorrow, wreck, and blight;
Think rather of those moments bright
When to the unconsciousness of right
    His course was true,
When Wisdom prospered in his sight
    And Virtue grew.

Yes, freely let our hands expand,
Freely as in youth's season bland,
When side by side, his Book in hand,
    We wont to stray,
Our pleasure varying at command
    Of each sweet Lay.

How oft inspired must he have trode
These pathways, yon far-stretching road!
There lurks his home; in that Abode,
    With mirth elate,
Or in his nobly pensive mood,
    The Rustic sate.

Proud thoughts that Image overawes,
Before it humbly let us pause,
And ask of Nature, from which cause
    And by what rules
She trained her BURNS to win applause
    That shames the Schools.

Through busiest street and loneliest glen
Are felt the flashes of his pen :
He rules, 'mid winter snows, and when
    Bees fill their hives :
Deep in the general heart of men
    His power survives.

What need of fields in some far clime
Where Heroes, Sages, Bards sublime,
And all that fetched the flowing rhyme
    From genuine springs,
Shall dwell together till old Time
    Folds up his wings ?

Sweet Mercy ! to the gates of Heaven
The minstrel lead, his sins forgiven ;
The rueful conflict, the heart riven
    With vain endeavour,
And memory of Earth's bitter leaven,
    Effaced for ever.

But why to him confine the prayer,
When kindred thoughts and yearnings bear
On the frail heart the purest share
    With all that live ?—
The best of what we do and are,
    Just God forgive !

To the Sons of Burns
After Visiting the Grave of their Father.

WILLIAM WORDSWORTH.

'MID crowded obelisks and urns
I sought the untimely grave of BURNS;
Sons of the Bard, my heart still mourns
    With sorrow true:
And more would grieve, but that it turns
    Trembling to you.

Through twilight shades of good and ill
Ye now are panting up life's hill,
And more than common strength and skill
    Must ye display
If ye would give the better will
    Its lawful sway.

Hath Nature strung your nerves to bear
Intemperance with less harm, beware!
But if the poet's wit ye share,
    Like him can speed
The social hour—for tenfold care
    There will be need.

Even honest men delight will take
To spare your failings for his sake,

Will flatter you,—and fool and rake,
  Your steps pursue:
And of your father's name will make
  A snare for you.

Far from their noisy haunts retire,
And add your voices to the quire
That sanctify the cottage fire
  With service meet;
There seek the genius of your sire,—
  His spirit greet:

Or where, 'mid "lonely heights and hows"
He paid to Nature tuneful vows;
Or wiped his honourable brows
  Bedewed with toil,
While reapers strove, or busy ploughs
  Upturned the soil;

His judgment with benignant ray
Shall guide, his fancy cheer, your way;
But ne'er to a seductive lay
  Let faith be given;
Nor deem that "light which leads astray,
  Is light from heaven."

Let no mean hope your souls enslave;
Be independent, generous, brave;
Your father such example gave,
  And such revere;
But be admonished by his grave,
  And think and fear!

Mossgiel.

William Wordsworth.

"There," said a stripling, pointing with meet pride
   Towards a low roof with green trees half concealed,
"Is Mossgiel farm; and that's the very field
   Where Burns ploughed up the daisy." Far and wide
A plain below stretched sea-ward, while descried
Above sea-clouds, the Peaks of Arran rose;
And by that simple notice the repose
Of earth, sky, sea, and air was vivified
Beneath "the random *bield* of cloud or stone"
Myriads of daisies have strove forth in flower
Near the lark's nest, and in their natural hour
Have passed away, less happy than the one
That by the unwilling ploughshare died to prove
The tender charm of poetry and love.

Burns at Mossgiel.*

CHARLES KENT.

BRIGHT dews of labour on his brow,
Warm passion in the ruddy glow,
Deep-flushing lustrous eyes below—
    What love flames back
Where thro' green leaves the white gleams flow
    That mark *her* track!

Sweet glimpse but of a rustic girl
With tartan veiled, whence streams one curl,
Where fluttering witcheries unfurl
    Love's springs of hair—
Of ringlets, yea! the pink, the pearl,
    His heart to snare!

Among the rippling wheat he stands,
A tawny reaper with brown hands,
That swathe ripe sheaves with crackling bands,
    Or with keen blade

---

* From "Dreamland, and other Poems." By Charles Kent. Longmans, 1862.

Sweep gold waves back from stubble-strands
   With shocks arrayed.

Rough, sunburnt, stalwart son of toil,
To till, to sow, to glean the soil,
How fair to thee that ringlet's coil
   That lures thy gaze !
Not rudest lot thy fame shall foil
   To chant her praise !

One moment there, one moment gone,
Quenchéd seems the arrowy beam that shone
That twinkling golden tress upon
   In trills of light—
Hope's shadowy mist of dreamings drawn
   Before thy sight !

Seen thro' which tremulous haze of hope,
Spread wide before thy fancy's scope—
As when o'er midnight's mystic cope
   God's gems are seen—
Strange visionary splendours ope
   And shine serene.

A young athletic peasant, thou !
Full soon Fame's crown shall gird thy brow
Thick gemmed with scarlet berries' glow,
   'Mid bristling leaves,
Thy sceptre, but a sickle now,
   Sway souls for sheaves.

That wondrous sceptre of thy song
Shall ever to thy land belong,
While every rapture, every wrong,
   That thrills thy breast,
By sympathy shall thrill the throng
   Thy woes have blest.

Then million millions yet unborn
Will hail with joy this autumn morn,
When loitering 'mid the ripened corn,
   Thy glorious eyes
Watched thro' yon maze of leaf and thorn
   Thy life's best prize.

Thy bonnie Jean, thy winsome wife,
Sweet blossom of that rugged life—
Rough rind with tenderest fibre rife,
   Whence bloomed yon flower,
Rich guerdon of thy manhood's strife,
   With healing power.

Was not her type that gowan fair,
When, toiling down the glebe of Ayr,
Thy footstep tracked the hissing share
   That turned the mould,
And pity yearned that jewel rare
   With love t' enfold?

The bonniest lass of blithest charms
Thou e'er didst win with wooing arms,

To soothe thee 'midst the world's alarms
　　In home's dear rest,
With looks whose merest memory warms
　　Thy manly breast.

The fairest of them all was she—
Yon "lass that made the bed for thee!"
To whom thy trust in grief may flee,
　　By anguish riven—
When Highland Mary e'en shall be
　　Still loved in heaven?

Unheard as yet Fame's trumpet-call
From yonder lowly labours' thrall
To grand Walhalla's deathless hall,
　　Where waits his throne—
Yon Peasant-Poet counts worth all
　　Her love alone!

Around him thus the day-beams shine
O'er locks more black than raven's crine,
O'er glittering orbs of light divine,
　　And radiant face,
Where sentience thrills each lordly line
　　With nerves of grace.

Ah! better, ROBIN, thus to stand
With sickle aye in healthful hand
Than leader of a brawling band
　　With gauge or bowl,
When bowed to sordid craft thy grand
　　Heroic soul!

Robert Burns.

JAMES MONTGOMERY.

WHAT bird, in beauty, flight, or song,
   Can with the Bard compare,
Who sang as sweet, and soar'd as strong,
   As every child of air?

His plume, his note, his form, could BURNS
   For whim or pleasure change;
He was not one, but all by turns,
   With transmigration strange.

The Blackbird, oracle of spring,
   When flowed his mortal lay;
The Swallow wheeling on the wing,
   Capriciously at play;

The Humming-bird, from bloom to bloom,
   Inhaling heavenly balm;
The Raven in the tempest's gloom;
   The Halcyon, in the calm;

In "Auld Kirk Alloway," the Owl,
   At witching time of night;
By "Bonnie Doon" the earliest Fowl
   That caroll'd to the light.

He was the Wren amidst the grove,
  When in his homely vein;
At Bannockburn the Bird of Jove,
  With thunder in his train;

The Woodlark in his mournful hours;
  The Goldfinch, in his mirth;
The Thrush, a spendthrift of his powers,
  Enrapturing heaven and earth;

The Swan in majesty and grace,
  Contemplative and still;
But roused,—no Falcon in the chase,
  Could like his satire kill.

The Linnet in simplicity,
  In tenderness the Dove;
But more than all beside was he—
  The Nightingale in love.

Oh! had he never stoop'd to shame,
  Nor lent a charm to vice,
How had Devotion loved to name
  That Bird of Paradise!

Peace to the dead!—In Scotia's choir
  Of minstrels great and small,
He sprang from his spontaneous fire,
  The Phœnix of them all.

For the Centenary of Robert Burns.

ROBERT LEIGHTON.

THE world is old ! States, Empires, Kings
   Have risen, ruled, and passed away;
Yet David harps, and Homer sings,
   And he of Avon speaks to-day.

The living song will still abide;
   And when our age is dust in urns
The world, as now, will own with pride
   Its life long debt to ROBERT BURNS.

His touch was universal birth;
   He set his native streams to tune;
And every corner of the earth
   Knows Nith and Lugar, Ayr and Doon.

His homes we seek, his haunts we trace
   Wherever thought of him is found,
We follow him from place to place
   And all is consecrated ground.

On things that disregarded lie
   His look bequeath'd a priceless dower,
The trodden daisy caught his eye
   And blossom'd an immortal flower.

Love's tender throes with him became
   A sweet religion ; and he poured
Such floods of beauty round a name
   That all men love whom he adored.

The patriot-hero's brows he bound
   With wreaths, eternal as the sun :
Tho' lowly honest man he crown'd ;
   He made the king and beggar one.

For well he knew that *lord*, or *king*,
   Was but a word. With deeper scan
He made both peer and peasant sing
   Their highest title still was—*man*.

In "shooting folly as it flew"
   There never was a deadlier aim ;
And even those his satire slew
   Are joint partakers of his fame.

He lashed the bigot ; his the creed
   Embracing all humanity ;
A conscience clear in word and deed—
   One Father, God ; and brethren, we.

And if we blame the sparkling rhymes
　That made the maddening cup sublime,
Think only of the alter'd times,
　And give the censure to the time.

In humour, friendship, pity, worth—
　In themes that change not with the day—
Broad Nature, felt o'er all the earth—
　His genius holds unmeasured sway.

Great Prince of Song! to mark thy fame,
　O for a moment of thy pen!
'Twere needless pains—thy living name
　Is written on the hearts of men.

Our gilt makes not thy gold more bright;
　But hearts enrich'd would yield returns;
A world of homage meets to-night,
　And every thought breathes ROBERT BURNS.

Address to the Shade of Burns.

Written for the Third Anniversary of the Irvine Burns Club, 1829.

CAPTAIN CHARLES GRAY.

HAIL BURNS! my native Bard, sublime;
Great master of our Doric rhyme!
Thy name shall last to latest time,
    And unborn ages
Shall listen to the magic chime
    Of thy enchanting pages!

Scarce had kind Nature given thee birth,
When from his caverns of the North,
Wild winter sent his tempests forth,
    The winds propelling—
To level with its native earth,
    The clay-built lowly dwelling.

Too well such storm did indicate
The gloom that hung upon thy fate;—
Arrived at manhood's wished estate,
    When ills were rife,

ADDRESS TO THE SHADE OF BURNS.

Thy heart would dance with joy elate
    At elemental strife !

Lone-seated by the roaring flood,
Or walking by the sheltered wood,
Rapt in devotion's solemn mood,
    Thy ardent mind
Left, whilst with generous thoughts it glowed,
    This sordid world behind !

Thou found man's sentence was to moil,
In turning o'er the stubborn soil ;
But ne'er was learning's midnight oil
    By thee consumed ;
Yet humour, fancy, cheered thy toil
    Whilst Nature round thee bloomed.

Though nurtured in the lowly shed—
A peasant born—with rustics bred—
Bright genius round thy head display'd
    Her beams intense—
Where Coila formed thee—loveliest maid !
    Ben i' the smeeky spence !

Mute is the voice of Coila now,
Who once with laurels decked thy brow ;—
Still let us ne'er forget that thou
    Taught learned men :
The hand that held the pond'rous plough
    Could wield the Poets' pen !

Upon thine eagle-course I gaze,
And weep o'er all thy devious ways ;
Tho' peer and peasant prized thy lays
    What did it serve ?
Grim av'rice said, "Give lasting bays,
    But let the Poet starve !"

The heartless mandate was obeyed ;—
Although the holly crowned thy head,
Yet wealth and power withheld their aid,
    And hugg'd their gain ;
While thy loved babes may cry for bread,
    And cry alas ! in vain !

But now *thy* column seeks the skies,
And draws the inquiring stranger's eyes ;—
Art's mimic boast for thee may rise
    Magnificent ;—
Yet thou hast reared, 'midst bitter sighs,
    A prouder monument !

Thy songs "untaught by rules of art,"
Came gushing from thy manly heart,
And claim for thee a high desert ;—
    In them we find
What genius only can impart—
    A mood for every mind !

The milkmaid at calm evening's close—
The ploughman starting from repose—
The lover weeping o'er his woes—
    The worst of pains !

The soldier as to fight he goes—
    All chaunt thy varied strains !

Sweet minstrel, "of the lowly train,"
We ne'er shall see thy like again !
May no rude hand thy laurels stain ;
    But o'er thy bier
Let poets breathe the soothing strain
    Through each revolving year !

Yes ! future bards shall pour the lay
To hail with joy thy natal day ;
And round thy head the verdant bay,
    Shall firm remain
Till Nature's handiworks decay,
    And chaos come again !

Robert Burns.

Dr. John M. Harper.

Written on the occasion of the poet's anniversary, and read before the Literary and Historical Society of Quebec, of which the author is Vice-President.

SWEET in the ear of Fame of yore a bard,
   With lips a lover's, wooed the heart of Time ;
To him his love alone was meet reward,
   Ere fame awoke to find his song sublime :
Within his heart the sheen of Nature glowed ;
A patriot's fire his noble soul endowed,
   And heart and soul found ecstacy in rhyme
That stirred the heart of time and soul of fame
To garland with the loves of men the Poet's name.

'Twas where the landscape sighs, where *Bonnie Doon*
   Sings mournfully as winter stays its glee,
The cottar's hearth, in light of Januar's moon,
   First heard the cry disguised of heaven's decree—
A Scottish poet born ; the north-wind blew
A trumpet-blast, but none the omen knew,
   Though drear the willows sighed across the lea,
And every sombre pine and bearded oak
Sustained the solemn strain till day awoke.

And day by day awoke till summer came,
   And year by year the rills renewed their song,
And year by year amid the sweet acclaim
   Of rural joys, the Poet's soul grew strong ;

For him ran clear the rhythm of the Doon;
For him its banks and braes were heaven's boon;
   When rang the glades with summer's warbling throng,
When blushing Nature laughed in every glen
To find her child, a poet, sing his sweet amen.

The poor maun thole what heroes whiles endure,
   Their night brings cheer through little hope of day;
And ROBIN's portion; not of shame, though poor,
   Distrained his thrift, the debt of fate to pay:
Yet while the glebe his labouring ploughshare gripped,
The cowering *Mouse* and *Daisy*, crimson-tipped,
   Less toilsome made for him life's weary way,
When love lit up the vista of its joys
With light, that from our soul's *Despondency* decoys.

To him was precious as the sweetest flower
   The incense of the *Cottar's* evening care—
A savour sweeter far than scented bower
   Of sanctity adorned at *Holy Fair:*
The altar of his father's faith was truth—
Truth born within, no outer silvered sooth
   Of *Unco Guid* or *Holy Willie's Prayer;*
And if from pleasure ROBIN stole a kiss,
The truth *A man's a Man—for a'* was surely his.

The scenes in amber gold of *Auld Langsyne*,
   The symphonies around our childhood's home,
The jewelled sward where browsed the sober kine,
   The hawthorn groves where love was wont to roam—
Of these he sang, and still his music thrills
The hearts of men to wile away their ills.

Even from his sadness suhbeams often come,
To foster in us wish to live again,
Where youth and mirth of yore began their wistful reign.

If reason frolics in the *Twa Dog's* name,
   And mense falls out near by the *Brigs of Ayr*,
If friendship's tryst neglects domestic claim,
   To jeopard prudence and the *Shanter's* mare ;
On other chords *John Anderson my Jo*
Plays sweet to soothe life's weary steps of woe,
   While *Man was made to mourn* makes men repair
The breach of fate, and through their grief hath raised
The test, that finds the good's in good and ill appraised.

If satire romps with *Hornbook* and his kind
   To tear the tinsel veil from falsehood's face,
If fraility dares the *De'il* and every hind
   Who seeks to drag the human through disgrace,
Love's sadness hath to joy a sweetness given
In *Highland Mary* and the hope that's heaven,
   While *Mailie Dead* finds elegy a place
Between our smiles and tears, as *Hallowe'en*,
With fun that fears, crowns sacred things with evergreen.

And when his wooing took heroic flight,
   His fervent spirit revelled in the past,
To sing the deeds of men who knew the right,
   And, knowing, dared maintain it to the last :
The thrilling, throbbing strains of *Scots Wha Hae*
Reverberate down the aisles of liberty :
   In them the pæan-peals of war outlast,

Though hushed, the torture of the patriot's task,
Though Scotia's glens, in sunshine born of peace, now bask.

'Twas thus the heart of time the Poet fanned,
   Thus won he claim to wear the *Vision's* wreath ;
Bred to the plough, his fame in every land
   Is scented with the fragrance of the heath :
The meadows fling his praises to the breeze,
The storm-winds echo them beyond the seas,
   And with them other bards bedew their faith,
Till every isle that loves the Saxon tongue
Hath with his lowland melody the welkin rung.

And Scotia's sons with patriotic cheer
   Join festival to celebrate his birth ;
The spirit of his song still hovers near,
   To lustre friendship and its well-timed mirth.
His song was Nature's—incense of the heart,
With naught to hide, because it knew no art,—
   The song of life as life is found on earth—
Sweetness in sorrow, evil out of good,
The only song man sings and yet hath understood.

How oft his minstrelsy entints our joys !
   How oft his genius bindeth friends sincere !
If life and joy we know he but alloys,
   'Tis these his love and poesy endear :
Hail to the land whose poet son he was !
Hail to the land that fought in freedom's cause !
   Hail to its love of song that runneth clear ;
With hand in hand as brethren let us seek
The virtue void of art, the patriot-pride that's meek.

Rantin' Robin, Rhymin' Robin.

David Vedder.

When Januar' winds were ravin' wil'
O'er a' the districts o' our isle,
There was a callant born in Kyle,
    And he was christen'd Robin.
Oh Robin was a dainty lad,
    Rantin' Robin, rhymin' Robin;
It made the gossips unco glad
    To hear the cheep o' Robin.

That ne'er to be forgotten morn,
When Coila's darling son was born,
Auld Scotland on his stock—an' horn,
    Play'd " Welcome Hame" to Robin.
And Robin was the blithest loon,
    Rantin Robin, rhymin' Robin,
That ever sang beneath the moon—
    We'll a' be proud o' Robin.

Fame stappin' in ayont the hearth—
Cried " I forsee your matchless worth,

And to the utmost ends o' earth
   I'll be your herald, ROBIN!"
And well she did emblaze his name,
   Rantin' ROBIN, rhymin' ROBIN,
In characters o' livin' flame,—
   We'll a' be proud o' ROBIN.

The Muses round his cradle hung,
The Graces wat his infant tongue,
And Independence wi' a' rung,
   Cried—"Redd the gate for ROBIN!"
For ROBIN's soul-arousing tones,
   Rantin' ROBIN, rhymin' ROBIN,
Gar'd tyrants tremble on their thrones,—
   We'll a' be proud o' ROBIN.

Then let's devote this night to mirth,
And celebrate our poet's birth;
While Freedom preaches i' the earth,
   She'll tak' her text frae ROBIN.
Oh! ROBIN's magic notes shall ring,
   Rantin' ROBIN, rhymin' ROBIN,
While rivers run and flowerets spring,
   Huzza! huzza for ROBIN!!

Ellisland.

PROF. JOHN STUART BLACKIE.

FAIR Ellisland, thou dearest spot,
On Scottish soil to each true Scot,
With wood and stream, and shining cot,
 Thy beauty sways me,
And love is rash—O blame me not
 If I shall praise thee!

Wide waves the leafy June around,
The banks with blossomy curls are crowned,
Sweet flows with mild and murmurous sound
 The clear Nith River,
And peace holds all the grassy ground
 Now sacred ever.

The Poet's farm! a fairer sight
Ne'er filled my view with calm delight;
Full fitly here our minstrel wight
 Did pitch his dwelling,
With beauty's green and gentle might
 Around him swelling!

Here stands the house, the very wall
Stout labour raised at ROBIN's call,
A farmer's beild, which, low and small,
 No envy breedeth,
Enough for comfort and for all
 A poet needeth.

And there the stack-yard where he lay
And gazed upon the starry ray,
When pensive Memory's tender sway,
    With fingers fairy,
Struck from his heart the sad sweet lay
    Of Highland Mary!

And here the bank where he did sit,
When once his quick and glancing wit
Off-started on a racing fit
    With glorious canter,
And forth with flashing hit on hit
    Flew Tam O'Shanter!

And oft, I ween, to that green bower
He walked, in placid evening hour,
With bonnie Jean, whose smile had power
    To soothe his spirit
When fitful thoughts, and fancies sour,
    Might rudely stir it!

Fair Ellisland! thou dearest spot
To each true-hearted stalwart Scot,
When I forget thy small white cot
    And winding river,
Sheer from my thought may Memory blot
    All trace for ever.

Burns' Birthday.

THOMAS BUCHANAN READ.

My friends, the grape that charms the cup to-night,
    Should be the noblest ever grown in cluster;
Our flowers of wit and song should be so bright,
    That all the place should wear a noon-tide lustre.

For he whose mortal day, and marvellous worth,
    We strive to honour with our yearly presence,
Was of that clay so seldom found on earth,
    On which the gods bestow their purest essence.

Ay, doubly bright should this ovation be;
    For we are honoured far beyond your dreaming;
The inward spirit bids me look and see,
    Where comes the bard with light and music teeming.

He comes, but not like Hamlet's sire, to wing
    The soul with fear, and urge to painful duty;
He comes; let us behold the phantom king—
    The king of song, and marvel at his beauty.

I see his presence in the luminous air,
    And feel no thrill to make my blood run colder;
He stands beside our presidential chair,
    With loving arm upon a Scotchman's shoulder.

Upon his brow a crown of glory beams ;
    His robe of splendour makes the lamplight hazy ;
In his right hand a pledging goblet gleams,
    The other holds a " crimson tipped daisy."

Of deathless rainbows is his tartan plaid ;
    His bonnet now is the celestial laurel ;
And on his face the light of song betrayed,
    Makes all the room with poesy grow choral.

With eye of inspiration stands the bard ;
    His lips are moving, though no sound can follow.
Let me translate,—although the task is hard,—
    To justly render Scotland's sweet Apollo.

" Dear friends, and brother Scotsmen doubly dear,"
    'Tis thus the poet looks his kind oration,
" The day is come, which once in every year
    Calls me to make my wonted visitation.

" I glide through Caledonian halls of mirth,
    Where votive feast and song together mingle ;
I seek the cot,—the sweetest place on earth
    Is just the simple peasant's glowing ingle.

" The haughty Briton lights his dusk saloon,
    Forgetting all his rancour for Prince Charlie,
And to the ploughman bard of Ayr and Doon,
    Pledges the smoking bree of Scottish barley.

" Where'er a ship upon the ocean swings,
    To-night, before the mariners seek their pillows,
My songs shall sail on their melodious wings,
    Like sea-birds o'er the phosphorescent billows.

"By Indian river, and Australian mine,
    And by the wall of China's old dominions,
My verse above their cups of mellow wine,
    Shall fan the air to music with its pinions.

"The far Canadian winter hears my name,
    E'en where the trapper's northern home is chosen,
The songs of Scotland, mingling with the flame,
    Warm all within, though all without be frozen.

"By Californian shores and forests old,
    Where, like a mighty bard new realms discerning,
The gray Pacific, over sands of gold,
    Chants his great song, the glittering metal spurning.

"In new-built towns, and round the miner's lamp,
    Or on the plains, or by the Colorado;
Where'er the far adventurous train may camp,
    My song to-night shall cheer the deepest shadow.

"Or in the snow-beleagured tents of strife,
    By jocund fires, or beds of painful story,
Health shall take courage, and the sick new life,
    To hear of Wallace, and of Bruce's glory.

"Oh, that my song may be as bolts of fire,
    Within the grasp of soldiers and of seamen!
The bard profanely wakes the sacred lyre,
    Who chants no strain to nerve the hearts of freemen.

"From town to town, obedient to the call,
    I pass in haste, for envious Time is fleeting,
As oft before, within this noble hall,
    I greet the friends who cheer me with their greeting.

"Here in your midst, my brothers, once again,
    I stand to-night a saddened guest and speaker;
I miss among you certain noble men,
    Who erewhile pledged me in a brimming beaker.

"For your sakes saddened,—not, my friends, for mine,—
    You mourn their music, and their pleasant sallies;
But we together pledge nectarean wine,
    And join our song in amaranthine valleys.

"I see the forms your sight cannot discern—
    I see the smile across their happy faces;
With eye of loving faith look round and learn
    Your friends are here,—there are no empty places.

"From shadowy goblets held in fingers dim,
    We drain the glass that keeps the memory vernal,
Our cups with yours are clinking brim to brim,
    And thus we pledge you in a draught fraternal.

"Adieu, adieu! across the eternal sea
    Still let us hear your pleasant song and laughter,
And let the love you bear me, warrant be,
    Of love as deep for all true bards hereafter."

Lines

Written on the Anniversary of Burns' Birth-Day, when wandering belated in the mountain passes on the frontier of Vermont.

HUGH AINSLIE.

WHEN last my feeble voice I raised
   To thy immortal dwelling,
The flame of friendship round me blazed,
   On breath of rapture swelling!

Now far into a foreign land,
   The heav'ns above me scowling,
The big bough waving like the wand,
   The forest caverns howling;

No kindred voice is in mine ear,
   No heart with mine is beating;
No tender eye of blue is near,
   My glance of kindness meeting;

But woody mountains, towering rude,
   Dare heaven with their statures !
'Tis Nature in her roughest mood,
   Amidst her roughest features !

Yet thou, who sang'st of Nature's charms,
   In barrenness and blossom,
Thy strain of love and freedom warms
   The chill that's in my bosom.

And here, where tyranny is mute,
   And right hath the ascendance,
O ! where's the soil could better suit
   Thy hymn of independence?

Thou giant 'mong the mighty dead !
   Full bowls to thee are flowing;
High souls of Scotia's noble breed
   With pride his might are glowing !

Burns.

JOHN IMLAH.

PRAISE to the poet's name who breathed
   On Scotia's ear the sweetest lays!
Hail to his natal day who wreathed
   The harp with greenest bays!
Was ever name so loved as his
   That o'er the Scottish heart so yearns?
Was ever day so dear as this
   That bore us ROBERT BURNS?

Yes! men and minstrels first among
   Is he whose name we honour now,—
Old Coila's son—the chief of song,
   The poet of the plough!
From castle hall to cottage hearth,
   Shall Scotia,—while this day returns
That gave her master minstrel birth,—
   Remember ROBERT BURNS!

Who breathed like him the burning strain
   Of lovers' fervour, hopes, and fears?

So knew the Muse's varied vein
   Of transport and of tears?
Or, if to rouse the patriot's soul,—
   The spirit that oppression spurns,—
Even to the death to glory's goal,—
   Who woke the lay like BURNS?

The wood-lark warbling on the spray,
   The daisy flowering at his feet,
Gave inspiration to his lay,
   Solemn and sad, yet sweet;
The homely feast of Hallowe'en—
   The ancient rites that Science scorns—
The pastimes of old days have been
   Embalm'd by ROBERT BURNS!

"The op'ning gowan wat wi' dew,"
   He twined with beauteous thought and theme,
The humblest bud the green earth grew
   His song has made supreme:
Ayr, Irvine, Lugar, Doon, and Nith,
   Through hazels, birks, or broom, or ferns,
Gleam in hallow'd glory with
   The deathless songs of BURNS!

The shepherd in his lonely shiel,
   The ploughman o'er the furrow'd field,
The maiden at her busy wheel,
   The cottar in his bield,
Have found a language in his lay
   Affection loves and memory learns—
The thoughts and feelings, grave or gay,
   Of Nature and of BURNS!

'Mid Western forests wide and drear,
  On lands beneath the burning line,
Sweet come upon the exile's ear
  The songs of "auld langsyne";
How fancy to the "banks and braes"
  Of early youth enrapt returns,
And lives o'er long departed days,
  Charm'd by the songs of BURNS!

Not narrow'd to his native spot,
  His soul embraced all Nature's plan,
He that knits Scot with brother Scot
  Binds man with fellow-man;
His harp the heart-strings of mankind,
  Each feeling knew his touch by turns,
And own'd the master hand and mind
  Of genius and of BURNS!

Wreathe laurels round the warrior's name,
  With thousands' tears and blood imbued,
Rear trophies to the monarch's fame
  For whom the sword subdued;
But time will hush the hireling's praise,
  The pile where marbled sorrow mourns—
The pyramid of future days
  Is raised to ROBERT BURNS!

For ever cherish'd be his name
  To whom the priceless gift was given,
High inspiration's holiest flame—
  The light that comes from heaven!
Praise to the child.—the chief of song,
  And may, as monumental urns,
All hearts bear on them deep and strong
  The memory of BURNS!

Robin's Awa!

JAMES HOGG.

As night i' the gloaming, as late I pass'd by,
A lassie sang sweet as she milkit her kye,
An' this was her sang, while her tears down did fa'—
O there's nae bard o' Nature sin' ROBIN's awa !
The bards o' our country, now sing as they may,
The best o' their ditties but maks my heart wae ;
For at the blithe strain there was ane beat them a'—
O there's nae bard o' Nature sin' ROBIN's awa !

Auld Wat he is wily and pleases us fine,
Wi' his lang-nebbit tales an' his ferlies lang-syne ;
Young Jack is a dreamer, Will sings like a craw,
An' Davie an' Delta are dowie an' slaw ;
Trig Tam frae the Hielands was ance a braw man ;
Poor Jamie he blunders an' sings as he can ;
There's the Clerk an' the Sodger, the News-man an' a',
But they gar me greet sairer for him that's awa !

'Twas he that could charm wi' the wauff o' his tongue,
Could rouse up the auld an' enliven the young,
An' cheer the blithe hearts in the cot an' the ha',
O there's nae bard o' Nature sin' ROBIN's awa !

ROBIN'S AWA!

Nae sangster amang us has half o' his art,
There was nae fonder lover, an' nae kinder heart;
Then wae to the wight wha wad wince at a flaw,
To tarnish the honours of him that's awa!

If he had some fauts I could never them see,
They're nae to be sung by sic gilpies as me,
He likit us weel, an' we likit him a'—
O there's nae sican callan sin' ROBIN's awa!
Whene'er I sing late at the milkin' my kye,
I look up to heaven an' say with a sigh,
Although he's now gane, he was king o' them a'—
Ah! there's nae bard o' Nature sin' ROBIN's awa!

Ode.

Written for, and read at the Celebration of Robert Burns'
Birth-day, Paisley, 29th January, 1805.

ROBERT TANNAHILL.

ONCE on a time, almighty Jove
Invited all the minor gods above,
   To spend one day in social festive pleasure ;
His regal robes were laid aside,
His crown, his sceptre, and his pride :
And wing'd with joy,
The hours did fly,
   The happiest ever time did measure.

Of love and social harmony they sung,
Till heav'n's high golden arches echoing rung ;
And as they quaff'd the nectar-flowing can,
Their toast was,
" Universal peace 'twixt man and man."
Their godship's eyes beam'd gladness with the wish,
And Mars half-redden'd with a guilty blush ;
Jove swore he'd hurl each rascal to perdition,
Who'd dare deface his works with wild ambition ;
But poured encomiums on each patriot band,
Who hating conquest guard their native land.

Loud thund'ring plaudits shook the bright abodes,
  Till Mercury, solemn-voic'd, assail'd their ears,
  Informing, that a stranger, all in tears,
Weeping, implor'd an audience of the gods.

Jove, ever-prone to succour the distrest,
A swell redressive glow'd within his breast,
He pitied much the stranger's sad condition,
And order'd his immediate admission.

The stranger enter'd, bowed respect to all,
Respectful silence reign'd throughout the hall.
His chequer'd robes excited their surprise,
Richly transvers'd with various glowing dyes :
A target on his strong left arm he bore,
Broad as the shield the mighty Fingal wore,
The glowing landscape on its centre shin'd,
And massy thistles round the borders twin'd ;
His brows were bound with yellow-blossom'd broom,
Green birch and roses blending in perfume ;
His eyes beam'd honour, tho' all red with grief,
And thus heaven's King spake comfort to the Chief.
" My son, let speech unfold thy cause of woe,
Say, why does melancholy cloud thy brow ?
'Tis mine the wrongs of virtue to redress ;
Speak, for 'tis mine to succour deep distress."
Then thus he spake : " O king ! by thy command,
I am the guardian of that far-fam'd land,
Nam'd Caledonia, great in art and arms,
And every worth that social fondness charms,
With every virtue that the heart approve,
Warm in their friendships, rapt'rous in their loves,
Profusely generous, obstinately just,
Inflexible as death their vows of trust :

For independence fires their noble minds,
Scorning deceit, as gods do scorn the fiends.
But what avail the virtues of the North,
No Patriot Bard to celebrate their worth,
No heav'n-taught Minstrel, with the voice of song,
To hymn their deeds, and make their names live long?
And, ah! should luxury, with soft winning wiles,
Spread her contagion o'er my subject-isles,
My hardy sons, no longer valour's boast,
Would sink, despis'd, their wonted greatness lost.
Forgive my wish, O king! I speak with awe,
Thy will is fate, thy word is sovereign law!
O, wouldst thou deign thy suppliant to regard,
And grant my country one true Patriot Bard,
My sons would glory in the blessing given,
And virtuous deeds spring from the gift of heaven!"
To which the god—"My son, cease to deplore,
Thy name in song shall sound the world all o'er;
Thy bard shall rise full-fraught with all the fire,
That heav'n and free-born nature can inspire:
Ye sacred Nine, your golden harps prepare,
T' instruct the fav'rite of my special care,
That whether the song be rais'd to war or love,
His soul-wing'd strains may equal those above.
Now faithful to thy trust, from sorrow free,
Go wait the issue of our high decree."
Speechless the Genius stood, in glad surprise
Adoring gratitude beam'd in his eyes;
The promis'd Bard, his soul with transport fills,
And light with joy he sought his native hills.

'Twas in regard of Wallace and his worth,
Jove honour'd Coila with his birth,

ODE.

       And on that morn,
       When BURNS was born,
       Each Muse with joy,
       Did hail the boy;
And fame on tip-toe, fain would blow her horn,
But Fate forbade the blast, too premature,
Till worth should sanction it beyond the critic's pow'r.

His merits proven—fame her blast hath blown,
Now Scotia's Bard o'er all the world is known—
But trembling doubts here check my unpolished lays,
What can they add to a whole world's praise;
Yet, while revolving time this day returns,
Let Scotchmen glory in the name of BURNS.

Robert Burns.

THOMAS FRASER.

KYLE claims his birth ;—wide earth, his name,
Where climes scarce kenn'd yet, peal his fame,
An' gaun time gayly chimes the same
    Where'er he turns,
Now, every true warm heart's the hame
    O' Minstrel BURNS !

Where Boreas brawls o'er blind'rin' snaw ;
Where simmer jinks through scented shaw ;
Where westlin' zephyrs saftly blaw,
    Their ROBIN reigns ;
An' even the thowless Esquimaux
    Hae heard his strains !

Dear Bonnie Doon, clear gurglin' Ayr,
Pure Afton an' the Lugar fair,
Can claim his sangs, their ain nae mair,
    Sin' lang years syne,
Braw Hudson an' thrang Delaware
    Kenn'd every line !

Frae zone to zone !—where'er we trace
The clearin' o' the pale-faced race ;—
Where still the red man trains the chase
　　　Through prairie brake,
E'en there his sang wi' sweet wild grace
　　　Rings round the lake !

The lone backwoodsman, as he seems
To ponder o'er his forest schemes,
Hums "Auld Lang Syne" among his dreams
　　　O' far-aff hame,
An' thinks, God bless him ! that the strains
　　　Croon ROBIN'S name !

Mothers wha skirled his sangs when bairns
In Carrick, Lothian, Merse or Mearns,
Are listenin' now by Indian cairns
　　　Wi' hearts half sobbin',
While some wee dawty blythely learns
　　　A verse frae ROBIN !

Sound though he sleeps in death's cauld bower,—
O ! what o' hearts this chosen hour,
Far as fleet fancy's wing can scower,
　　　In raptured thrangs,
Are thriling wi' the warlock power
　　　O' ROBIN'S sangs.

Frae Alloway's auld hunted aisle
To far Australia's gowd-strewn soil ;
And e'en where India's ruthless guile
　　　Mak's mercy quake,
Soul-minglin' there, worth, wealth and toil
　　　Meet for his sake.

True hearts at hame—true to the core,
To auld Scots bards an' auld warld lore,
Are blendin'—as in scenes o' yore,
    Wi' BURNS the van—
Love for braw Clydesdale's wild woods hoar
    An' love for man.

Staid Arthur's Seat's grim grey man's head
Bows to Auld Reekie's requiem reed;
While Soutra lifts the wailin' screed,
    An' Tweed returns
His plaintive praises o'er the dead,
    The darlin' BURNS.

Poor dowie Mauchline dights her e'e;
Nith maunders to the sabbin' sea;
An' high on Bannock's far-famed lea
    The stalwart thistle
Droops as the winds in mournfu' key
    Around him rustle.

Dark glooms Dumfries, as slowly past
Saunt Michael's growls the gruesome blast,
Where Scotia, pale an' sair down-cast,
    Clasps the sad grun'
That haps her loved, and to the last
    Immortal Son !

While backward frae the grave-yard drear,
Thought, tremblin' through a hundred year,
Sees Doon's clay cot, where weel hained cheer,
    Show's poortith's joy
When Nature's sel' brought hame her dear,
    Choice, noble boy.

But soon blythe hope fu' kindly keeks
Within her wae-sunk heart, an' seeks
To tint her trickling snaw-white cheeks
  Wi' words that burn,—
Why! when a world her Bard's fame speaks!
  Why should she mourn!

Wide though the great Atlantic rows
His huge waves, wi' their wild white pows,
To part our auld an' new warld knowes,
  Weel pleased, she turns
A westward look, where lustrous grows
  The name o' BURNS!

Pride, too, though tear-dimmed for a wee,
May lively light her heart wi' glee,
For where, sin' winged earth first flew free,
  E'er lived the lan'
That bore so true a Bard as he—
  So true a Man?

In him poor human nature's heart
Had ae firm friend to take its part,
So weel kenn'd he wi' what fell art
  Our passions goad
Frail man to slight fair virtue's chart,
  An' lose his road.

An' we, whose lot's to toil, an' thole,
Though cross an' care harass the soul,
Can cheer the weary wark-day's dole
  Wi' strains heart-wrung,
Brave strains! our BURNS, worn but heart-whole,
  Alone has sung.

His words hae gi'en truth wings, to bear
Round earth the poor man's faith, that here
Vain pride can ne'er wi' plain worth peer,
    Nor lift aught livin'
Ae foot, though tip-tae raxed on gear,
    The nearer heaven.

Fearless for right, wi' nerve to dare,
Seer-like he laid his sage soul bare,
To show what life had graven there,
    That earth might learn ;—
Yet, though a' earth in BURNS may share,
    He's *Scotia's* bairn !

An' O ! how dearly has he row'd
Her round wi' glory, like the gowd
Her ain braw sunset pours on cloud,
    Crag, strath an' river,
Till queen o' sang she stands, uncowed,
    An' crowned for ever !

Whilst we within our heart's-heart shrine
The man—"The brither man !"—entwine
Wi' a' the loves o' auld lang syne !
    An' young to-day,
Scotland an' BURNS !—twa names to shine,
    While Time grows grey !

Scotland hersel' !—wi' a' her glories,
Her daurin' deeds an' dear auld stories ;
The great an' guid, wha've gane before us ;
    Her martyr host ;
E'en wi' the graves o' them that bore us,
    The loved an' lost.

Her sword, that aye flashed first for right ;
Her word, that never craved to might ;
Her sang, brought down like gleams o' light
    On music's wings,
To nerve her in the lang fierce fight
    Wi' hostile kings.

Her laverock, in the dawnin' clouds ;
Her merle, amang the evenin' woods ;
Her mavis, 'mang the birk's young buds ;
    The blythe wee wren,
An ROBIN's namesake, as he scuds
    Through drift-white glen.

Her snawdrap, warslin wi' the sleet ;
Her primrose, pearled wi' dewy weet ;
Her bluebell, frae its mountain seat
    Beckin' an' bowin',
Her wee gem, sweetest o' the sweet,
    The peerless gowan.

Her waters, in their sangsome glee,
Gurglin' through cleuch and clover-lea,
Soughin' aneath the saughen tree
    Where fishers hide,
An' driftin' outward to the sea
    Wi' buirdly pride.

The catkins, that her hazels hing
In clusters round the nooks o' spring ;
Her rowan, an' her haws, that swing
    O'er wadeless streams,
An' bless the school-boy hearts, that bring
    Them hame in dreams.

Her muirlan's, in their heather bloom;
Her deep glens, in their silent gloom;
Her gray crags, where their torrents fume
    Wi' downward shiver;
Her braesides, wi' their thistle plume,
    Free, an' forever!

Scotland hersel'—Heaven bless her name!
Wi' a' her kith an' kin the same—
Yes! Scotland's sel', wi' a' her fame,
    Weel's we revere her,
Than him, her Bard o' heart an' hame,
    Is scarcely dearer!

So rare the sway, his heart-strains wield,
In lordly ha' an low thack bield,
Wi' manhood, youth an' hoar-browned eild,
    O'er Scotland wild,
BURNS an' The Word, frae Heaven revealed,
    Lie side by side.

Earth owned! his genius in its prime,
Now towers in mind's fair green-hilled clime,
Where, mist-robed, Ossian out-sings time,
    An' Shakespeare smiles,
As Milton, murmurin' dreams sublime,
    Looks earthward whiles!

O! hear then, Scot!—though yet you toil
To fill some lordlin's loof wi' spoil,
Or thriving on Columbian soil,
    Yoursel' your lord,
Ne'er dim his now bright fame wi' guile
In thought or word!

Spurn a' that's wraug, an' mak' the right
Your haudfast sure, stieve strong an' tight,
Cling there, an' ne'er let out of sight
    The wants o' man,
But BURNS-like, strive his lot to light
    As weel's you can.

Ne'er let vile self get grip, to twist
What heart or conscience dictates just ;
Straightforward aye act, though fate's gust
    May take your breath ;—
The man wha fears nae face o' dust,
    Needs scarce fear death.

Proud, stern, though gentle as the tone
Breathed through a mother's prayerfu' moan,
BURNS scorned to snool round rank or throne,
    Fause-tongued an' tame ;—
Till death, his heart was freedom's own ;
    Be ours the same !

The Bard of Song.
Written for Burns' Anniversary, 1834.

ROBERT GILFILLAN.

THE bard of song rose in the west,
  And gladdened Coila's land,
The badge of fame was on his brow,
  Her sceptre in his hand.

The minstrel Muse beheld her son,
  While glory round him shone,
Walk forth to kindle with his glance
  Whate'er he looked upon !

She saw the green earth where he strayed
  Acquire a greener hue,
And sunny skies high o'er his head
  Assume a brighter blue.

She saw him strike his rustic harp,
  In cadence wild and strong :
His song was of bold freedom's land—
  Of Scotland was his song !

He soared not 'mong aerial clouds,
  Beyond the mortal ken ;
His song was of the moorland wild,
  The happy homes of men.

Or of our battle chiefs, who rose
  To his enraptured view—
He knelt before the Bruce's crown,
  And sword that Wallace drew.

THE BARD OF SONG.

Their deeds inspired his martial strains,
  He marked the patriot band
Who stood, 'mid dark and stormy days,
  The guardians of our land.

"All hail! my son," the Muse she cried,
  "Thy star shall ne'er decline;
A deathless name, and lasting fame,
  Shall evermore be thine!"

Fain had she said, "and length of days,"
  But thus she boding sung—
"Away, away, nor longer stay,
  Thy parting knell hath rung!"

The Minstrel sighed, and from his harp
  A few sad tones here fell;
They told of honours—all too late,
  And of his last farewell!

They told of fame, when he no more
  Would need a cold world's fame—
Of proud memorials to his name,
  When he was but a name!—

Of pride, of contumely, and scorn—
  The proud man's passing by—
The Minstrel left to die on earth,
  Yet lauded to the sky!

'Tis past!—and yet there lives a voice
  That thrills the chords among:
'Tis—Scotland's song shall be of BURNS,
  Who gave to Scotland song.

Ode.

The Anniversary of Robert Burns, January 1815.

WILLIAM GLEN.

COME, my sweet harp, come murmur on,
   Sing of my home in glorious glee;
A fairer land than Caledon,
   Ne'er started from a stormy sea,
And fling thy numbers bold and free,
   To him whose notes roll'd sweet along,
For dear as life, as Heaven, will be
   The land of freedom and of song.

Let Haffiz live in Persian strains,
   Let Italy her Tasso claim,
Let Homer charm the Grecian plains,
   His country's boast, his country's shame;
Let Milton raise fair England's name,
   And genius consecrate their urns;
But where's the Bard can cloud the fame
   Of Scotland's pride, her darling BURNS?

Ye masters of the ancient school,
   Ye moderns wooing genius mild,

ODE.

Know that a Bard's not form'd by rule,
   Bright-polish'd till the fabric's spoil'd.
O! give me Nature's artless child,
   Who spurns all gaudy tinsel glare,
Like him who sung his "wood-notes wild"
   Upon the bonnie banks of Ayr.

"O! thou pale orb," thou'st seen him stray,
   By Nith's sweet winding lovely stream,
Giving bright fancy all its play,
   Whilst gazing on thy wandering beam;
Thou'st mark'd with sweet diffusing gleam,
   Him mourning by Lincluden towers,
"How life and love were all a dream,"
   And he confess'd their bitter powers.

Yet oft in merriment and glee,
   He "set the table in a roar,"
Wild as the wildest could he be,
   And ablest wits confess'd his powers;
Yet all at once could he restore
   The woe-tear to the eye again,
Bid mirth's mad witchery charm no more,
   And call to life sad sorrow's train.

Coila; thy vales are silent now,
   He's gone who all thy beauties drew,
Go bind on thy majestic brow,
   The weeping rosemary and rue;
And let the sorrow-shading yew,
   Hang o'er the grave where Nature mourns,
And weep, sweet Coila, for I trow
   You lost your brightest gem in BURNS.

ODE.

"While ruin's ploughshare drives elate,"
    While men their fellow-mortals spurn,
And weeping pleasure's transient date,
    Exclaims that "Man was made to mourn."
Or if from every rapture torn,
    We sadly wail a darling maid,
We'll know his wae who called forlorn
    On "Mary's dear departed shade."

Or when our fathers' deeds he grac'd,
    Raising their deathless fame on high,
Bade us while every wae be trac'd,
    Wail Scotland's fallen majesty;
Or brought the tear-drop in our eye,
    When resting on her lowly tomb,
And bade us heave the unconscious sigh,
    When mourning hapless Mary's doom.

Whether he struck the notes of woe,
    Or bade them with wild joy expand,—
In pleasure's tide, or sorrow's flow—
    His lyre was sweet, majestic, grand:
He touch'd it with a master's hand,
    Its heavenly tone will never die,
And many, many a distant land
    Was charmed with his minstrelsy.

We'll lay the lyre upon his urn,
    And while the moon-beams deck the plain,
Mayhap his spirit may return,
    And sweep the trembling chords again,
And we may hear the fairy strain,
    Float on the night-breeze down the dell—
Delusion all, it is in vain—
    And now, sweet Bard, again farewell.

Verses

Written on Visiting the House in which Robert Burns was Born, and the surrounding Scenery, in Autumn, 1799.

RICHARD GALL.

O BUT it makes my heart fu' sair,
    The lowly blast-worn bower to see,
Whare infant Genius wont to smile,
    Whare brightened first the Poet's e'e !

BURNS, heavenly Bard ! 'twas here thy mind
    Traced ilka object wildly grand ;
Here first thou caught dame Nature's fire,
    An' snatched the pencil from her hand.

Bleak Autumn now reigns o'er these scenes,
    The yellow leaves fa' aff the tree ;
But never shall the laurel fade,
    That Scotia's Muse has twined for thee.

O Doon ! aft wad he tent thy stream,
    Whan roaming near the flowery thorn,
An' sweetly sing " departed joys,
    Departed never to return ! "

An' near thy bonny crystal wave,
   Reft o' its rose we find the brier,
Beneath whase shade he wont to lean,
   An' press the cheek o' Jeanie dear.

O'er yonder heights, in simmer tide,
   His canty whistle aften rang;
An' this the bank, an' this the brae,
   That echoed back the Ploughman's sang.

But whare is now his wonted glee,
   That sic enchanting pleasure gave?
Ah me! cauld lies the Poet's head;
   The wintry blast howls o'er his grave!

To ither lands the Poet's gane,
   Frae which the traveller ne'er returns;
While Nature lilts a waefu' sang,
   And o'er her Shakspeare Scotia mourns.

Coila's Bard.

JAMES STIRRAT.

THERE's nae bard to charm us now,
   Nae bard ava
Can sing a sang to Nature true,
   Since Coila's bard's awa'.

The simple harp o' earlier days
   In silence slumbers now,
And modern art wi' tuneless lays
   Presumes the Nine to woo.
      But nae bard in a' our isle,
         Nae bard ava
      Frae pawky Coila wins a smile
         Since ROBIN gaed awa'.

His hamely style let Fashion spurn—
   She wants baith taste and skill;
And wiser should she ever turn,
   She'll sing his sangs hersel'.
      For nae sang sic pathos speaks,
         Nae sang ava;
      And Fashion's foreign rants and squeaks
         Should a' be drumm'd awa'.

Her far-fetch'd figures aye maun fail
　　To touch the feeling heart;
Simplicity's direct appeal
　　Excels sic learned art.
　　　　And nae modern minstrel's lay,
　　　　　　Nae lay ava,
　　　　Sae powerfully the heart can sway,
　　　　　　As ROBIN's that's awa'.

For o'er his numbers Coila's muse
　　A magic influence breathed,
And roun' her darling poet's brows
　　A peerless crown had wreathed.
　　　　And nae wreath that e'er was seen,
　　　　　　Nae wreath ava,
　　　　Will bloom sae lang's the holly green
　　　　　　O' ROBIN that's awa'.

Let Erin's minstrel, Tommy Moore,
　　His lyrics sweetly sing,
'Twad lend his harp a higher bower
　　Would Coila add a string.
　　　　For nae harp has yet been kent,
　　　　　　Nae harp ava,
　　　　To match the harp by Coila lent
　　　　　　To ROBIN that's awa'.

And though our shepherd, Jamie Hogg,
　　His pipe far sweetly plays,
It ne'er will charm auld Scotland's lug
　　Like ploughman ROBIN's lays.
　　　　For nae pipe will Jamie tune,
　　　　　　Nae pipe ava,
　　　　Like that which breathed by "bonnie Doon"
　　　　　　Ere ROBIN gaed awa'.

Even Scotland's pride, Sir Walter Scott,
  Wha boldly strikes the lyre,
Maun yield to ROBIN's sweet-love note,
  His native wit and fire.
    For nae bard hath ever sung,
      Nae bard ava,
    In hamely or in foreign tongue,
      Like ROBIN that's awa'.

Frae feeling heart Tom Campbell's lays
  In classic beauty flow,
But ROBIN's artless sangs displays
  The saul's impassioned glow.
    For nae bard by classic lore,
      Nae bard ava,
    Has thrill'd the bosom's utmost core
      Like ROBIN that's awa'.

A powerfu' harp did Byron sweep,
  But not wi' happy glee;
And though his tones were strong and deep,
  He ne'er could change the key.
    For nae bard beneath the lift,
      Nae bard ava,
    Wi' master skill the keys could shift
      Like ROBIN that's awa'.

He needs nae monumental stones
  To keep alive his fame,
Auld Granny Scotland and her weans
  Will ever sing his name.
    For nae name does fame record,
      Nae name ava,
    By Caledonia mair ador'd
      Than ROBIN's that's awa.

Elegy to the Memory of Robert Burns.

ALEXANDER BALFOUR.

THE lingering sun's last parting beam
   On mountain tops had died away,
And night, the friend of Fancy's dream,
   Stole o'er the fields in dusky grey ;

Tired of the busy, bustling throng,
   I wandered forth along the vale ;
To list the widowed blackbird's song,
   And breathe the balmy evening gale.

I leaned by Brothock's limpid tide,
   The green birch waving o'er my head ;
While night winds through the willows sighed,
   That wept above their watery bed ;

'Twas there the Muse without control,
   Essayed on fluttering wings to rise ;
When listless langour seized my soul,
   And drowsy slumbers sealed my eyes ;

In Morpheus' arms supinely laid,
   My vagrant Fancy roved astray,
When lo ! in radiant robes arrayed,
   A spirit winged its airy way.

In dumb surprise, and solemn awe,
   I wondering gazed, when by my side
A maid of matchless grace I saw,
   Arrayed in more than mortal pride;

Her eye was like the light'ning's gleam,
   That can through boundless space pervade,
But sorrow seemed to shade its beam,
   And pallid grief her cheeks o'erspread;

A flowery wreath, with bays entwined,
   Fresh blooming from her girdle hung;
And on the daisied bank reclined,
   She touched a Harp for sadness strung:

The trembling strings—the murmuring rill—
   The hollow breeze that breathed between—
Responsive echo from the hill—
   All joined to swell the solemn scene!

The maid, in accents sadly sweet,
   To sorrow gave unbounded sway;
My fluttering heart forgot to beat,
   While thus she poured the plaintive lay.

"I am the Muse of Caledon,
   From earliest ages aye admired;
Through her most distant corners known,
   Oft has my voice her sons inspired.

"My charms have fired a royal breast—
   A King who Scotia's sceptre bore—
I soothed his soul, with trouble pressed,
   When captive on a hostile shore:

"My bays have on a Soldier's brow,
  Amidst his verdant laurels twined;
Inspired his soul with martial glow,
  And called his country's wrongs to mind.

"The warblings of my Harp have won,
  A mitred son from Holy See;
Who oft from morn to setting sun,
  Would hold a Carnival with me:

"But chief of all the tuneful train,
  Was BURNS—my latest—fondest care!
I nursed him on his native plain;
  And now, his absence is—despair!

"I hailed his happy natal hour,
  And o'er his infant cradle hung;
O'er Fancy's wild, unbounded power,
  Or Reason's earliest bud was sprung.

"I saw the young ideas rise
  Successive in his youthful mind;
Nor could the peasant's garb disguise
  The kindling flame, that lay confined.

"Oft have I met him on the dale,
  Companion of the thoughtless throng;
And led him down the dewy vale,
  To carol o'er some artless song.

"Unseen by all, but him alone
  I cheered his labours through the day;
And when the rural task was done,
  We sought some wild sequestered way;

ELEGY.

" Midst Coila's hills, or woodlands wild,
　　By Stinchar's banks, or Lugar's stream,
There would I place my darling child,
　　And soothe him with some pleasing dream.

" These haunts to him were blissful bowers,
　　Where all the soul was unconfined ;
And Fancy culled her choicest flowers,
　　To warm her youthful poet's mind.

" Nursed on the healthful happy plains,
　　Where love's first blush from Virtue springs,
'Twas Nature taught the heartfelt strains,
　　That o'er the vassaled Cot he sings.

" Keen Poverty with withered arms,
　　Compressed him in her cold embrace ;
And mental griefs, ungracious harms
　　Had furrowed o'er his youthful face.

" Yet there, the dear delightful flame,
　　Which rules the breast with boundless sway :
Resistless fired his melting frame,
　　And taught the love-lamenting lay.

" A friend to Mirth, and foe to Care,
　　Yet formed to feel for worth oppressed ;
His sympathetic soul could share
　　The woes that wring a brother's breast.

" Ah ! gentle Bard ! thy tenderest tear,
　　Was o'er a hapless orphan shed !
But who shall thy sweet prattlers' cheer,
　　Now that a green-turf wraps thy head ?

"He who can still the raven's voice,
    And deck the lily's breast like snow,
Can make thy orphan train rejoice,
    And soothe thy widow's song of woe.

"Ye souls of sympathetic mind,
    When smiling Plenty deigns to crown,
Yours be the task their wounds to bind,
    And make their happiness your own.

"To banish Want, and pale-faced Care,
    To wipe the tear from Misery's eye,
To such a bliss as Angels share,
    And tell with joy above the sky!

"Where are the thrilling strains of woe
    That echoed o'er Glencairn's sad urn?
And where is now oppression's foe
    Who taught, that 'Man was made to mourn'?

"Why when his morning calmly smiled,
    Did Hope forebode a lengthened day?
My promised joys are now beguiled,
    Since darkness hides my darling's clay!

"Yet rest in peace, thou gentle shade!
    Although the 'narrow house' be thine;
No pious rite shall pass unpaid,
    No hands unhallowed stain thy shrine.

"The blighting breath of venomed Scorn
    Shall harmless round thy mansion rave;
Though Envy plan her poignant thorn,
    It ne'er shall bud above thy grave.

ELEGY.

"The stagnant soul, unmoved, may hear
    Of worth, if ne'er was formed to feel;
The selfish heart, with haughty sneer,
    Unblushing, boast a breast of steel:

"Yet Sympathy, that loves to sigh,
    And Pity, sweet celestial maid,
And Genius, with her eagle eye,
    Shall hover round thy hallowed shade.

"The torrent dashing down the steep,
    The wild wave foaming far below,
In Nature's notes for thee shall weep,
    With all the majesty of woe!

"When Winter howls across the plain,
    And spreads a thick obscuring gloom,
His winds on Coila shall complain,
    And hoarsely murmur o'er thy tomb!

"There, virgin Spring shall first be seen,
    To deck with flowers thy dewy bed;
And Summer, robed in richest green,
    Shall hang her roses o'er thy head.

"When Autumn calls thy fellow swains
    (Companions now, alas! no more!)
To 'reap the plenty of their plains,'
    Their mingling sighs shall thee deplore.

"O pour a tear of tenderest woe,
    Ye bards who boast congenial fire;
Let sympathetic wailings flow,
    And Sorrow's song attune the lyre.

" Ye warblers, flitting on the wind,
　　Chaunt forth your saddest plaintive strain;
And weep—(for ye have lost a friend),
　　Ye little wanderers of the plain !

" This garland, for my bard entwined,
　　No brow but his shall ever wear;
Around his turf these flowers I'll bind,
　　And wet them nightly with a tear !

" While dews descend upon his tomb,
　　So long the Muse shall love his name;
Nor shall this wreath forget to bloom,
　　Till latest ages sing his fame.

" But still, officious friends, beware !
　　Nor rashly wound my favourite's fame;
O watch it with parental care !
　　Stain not the hapless Minstrel's name.

" Seek not, amidst his wreath to twine
　　One verse that he himself suppressed;
His offerings made at folly's shrine,
　　Let them in dark oblivion rest !

" Ye wanderers in the wilds of song,
　　On whom I have not smiled in vain,
Would you the blissful hours prolong,
　　O shun seductive Pleasure's train !

" The bays that flourish round her bowers,
　　Are venomed o'er with noxious dews;
The thorns that lurk amidst her flowers,
　　A rankling poison oft infuse.

"Though Luxury's lap seem softly spread,
    The couch of joy, and sweet repose,
Yet hissing furies, haunt her bed,
    And rack the mind with keenest woes.

"The hedge-rowed plain, the flowery vale,
    Where rosy Health delighted roves,
Where Labour tells his jocund tale,
    And village maidens sing their loves,—

"'Tis there the Muse unfolds her charms;
    From thence her sons should never stray;
Ye souls whom boundless fancy warms,
    Still keep this calm sequestered way;

"So may such never-dying praise,
    As echoes o'er my darling's tomb,
Congenial bloom amidst your bays,
    And Heaven bestow a happier doom!"

She ceased her song of sorrow deep,
    Her warbling harp was heard no more;
I waked—and wished again to sleep—
    But ah! the pleasing dream was o'er.

The rustic Muse, untaught to sing,
    Has marred the Vision's solemn strain;
Too harshly touched the pensive string,
    To soothe thy shade, lamented swain!

Unskilled to frame the venal lay
    That flows not from a heart sincere,
'Tis mine this artless meed to pay—
    The heart-felt sigh—and silent tear.

What is Success?

OR

THE PHILOSOPHER AND THE PLOUGHMAN.

Thomas C. Latto.

'Mong the pleasant shades of Catrine,
   Calm, dignified and cool,
In his metaphysics buried,
   Set a sage of the old school.
Out roll'd long, ponderous periods,
   That fair hand scorn'd to write,
Which a deft amanuensis *
   Sat down in black and white.

Aristotelian systems,
   Vast speculations deep,
Despair of learnèd pundits,
   Traversed his mental sweep;
Condorcet's fine-spun theories,
   Be sure he did not spare;
Descartes' soap-bubble vortices
   He blew into the air.

---

* Hew Ainslie, author of "The Ingle Side," "The Rover of Loch Ryan," etc.

WHAT IS SUCCESS?

All passions and emotions—
   Hate, joy, revenge and love—
He counts their pulse-beats coolly,
   Through the fingers of his glove;
Of "Free Will" and "Necessity,"
   "No" and "Eternal Yea,"
Like Solomon he reasons
   Through the long summer day.

Perchance to find the siftings
   Through shadowy hopper pass'd,
Like the fine dust of "Willie's Mill,"*
   But vanity at last.
Nay! no such doubts disturb'd him.
   In his complacent ease;
His gnats were stately elephants;
   Fat lobsters were his fleas.

Leibnitz' "monads," quite adroitly,
   He turns upon his fork;
Hartley's "vibratiuncles"
   Dissects like tender pork;
Voltaire is sharply riddled;
   E'en Mallebranche gets a hoist,
And grim old Hobbes of Malmesbury
   Hangs dangling from a joist.

For him Encyclopædias
   Their tentacles stretch'd out,
To feel what feat in Ethics
   Was *Atticus* about.

* See "Tam o' Shanter."

WHAT IS SUCCESS?

While thro' the press no Dominie
   His last grand sermon ran,
Without the *Imprimatur*
   Of our august *savan*.

He dreamt of an inheritance
   Held by a tenure sure,
Among the Oracles of Eld
   For ever to endure,
While flourished High Philosophy
   And Plato's lore the theme,
The far-famed Catrine groves would bloom,
   A second Academe.

There, in luxurious library,
   Curtain'd with rich brocade,
His traps for immortality,
   The bland Professor laid,
'Twere heartless to estop what some
   A wild-goose chase might deem;
Still, let the venerable sage
   Indulge his waking dream.

O Dignity! what splendours
   Wait thy supreme behest!
Why rolled not *I* in purple
   With swelling, princely crest?
Wealth, honours, in the vestibule,
   Obsequious, bow the knee;
What more needs man that's mortal
   Than genial smile from thee?

WHAT IS SUCCESS?

WITHIN a hovel, far away,
    In a cramp'd attic loft,
A swarthy lad, with great black eyes,
    Imperiously soft ;
With horny hands, as daily used
    To cleave the stubborn soil,
Scribbled in moments snatch'd by stealth
    From his accustomed toil.

No stenographic aide-de-camp
    Looked up to him as lord,
With sharpen'd quill to get his cue
    Or prompt the lacking word.
No desk—that were appliance strange
    In a bare, dingy nook—
No roll of crabbed manuscript,
    Not e'en a tatter'd book.

Stern, sometimes savage, was his glance,
    As if he satire penn'd ;
Anon his dark eye beam'd with love
    That few could comprehend.
This too, would pass and in its place
    Resolve and dauntless grit,
Stalks of the " carle hemp " in man
    Upon his brow were writ.

Deep glow'd in those great wondrous orbs
    The fire of genius true,
As lines that ages yet would prize
    From broad-nibb'd " stumpie " flew.
O'er the rude table of fir deal,
    He lean'd and wrote a while,
Then in the drawer the scrolls were flung
    With half sarcastic smile.

Up with his gad and off to work;
  His "fittie-lan "* that shared
The forenoon's darg, her oats well munched,
  Is for fresh stent prepared.
But hark! a light foot on the trap
  Springs eager as a bird;
A young lass, drawing forth a script,
  Devours it every word.

His only sister! who but she
  Who knows she there will find
The graphic pictures, genius-born,
  Sparks of a noble mind.
Entranced, an hour she whiles away;
  With pride her heart rebounds,
Till summoned by the mother's voice
  To her domestic rounds.

There is a debt that presses sore,
  Enquire not of the cause,
And debts have never yet been paid
  By country-folks' applause.
The songs appear in "gude black prent,"
  " Wee Johnnie "† waves his wand;
Within a month that ploughman lad
  Towers foremost of the land.

To fair Edina's shrine he wends
  Odd mart to earn his meal;—
And—as is meet—the letter'd sage
  Must see the "ploughman chiel."

---

* The nearer horse of the hendmost pair at the plough—the leader.
† John Wilson, Printer, Kilmarnock.

WHAT IS SUCCESS?

The " braid claith " meets the "hodden gray "
  And, not incurious, views
The hob-nailed *lion* of the day—
  Last item of the " News."

With glance at the bucolic youth,
  Brimful of cautions wise,
As Zeus might on a linnet look,
  With calm majestic eyes.
The Doctor condescends advice,
  With mild Mecenas air,
To him of whom the peopled globe
  Has now assumed the care.

Then creeps the Bard back to his den,
  To feel and know, in sooth,
What bitterness 'tis to bemoan
  A sad unfriended youth ;—
The cup dashed from his longing lips,
  That, in rare bours of glee,
When " rantin' round in Pleasure's ring "
  He never thought to " pree."

Now, where the sage ? the rustic, where ?
  Strangely their lot has changed,
Since by the furzy braes of Braid
  At morning hour they ranged.
For recognition only marks
  The philosophic name,
Because it chanced to cross the disk
  Of BURNS' immortal fame.

Burns.

William Murray.

Illustrious poets lived and sang before
   King Robin warbles. Stirring strains in streams,
From his own land alone, subdued the roar
   Of bigot's ban and sacerdotal screams.
Before and after him, old Scotland's glens
   Have echoed with successions of sweet song,
Which Scotland's rocks and rivers, braes and bens,
   In grateful sympathy and love prolong.
But Burns, ablaze in wild Creation's eye,
   Outshines all rivals, Jove-like in his mould,
He glows with equal splendour far or nigh,
   And still will radiate when Creation's cold.
The more the marvel of his Muse we measure,
The more he fills us with perpetual pleasure.

The Birthplace of Robert Burns.

ROBERT G. INGERSOLL.

Though Scotland boasts a thousand names
    Of patriot, king, and peer,
The noblest, grandest of them all
    Was loved and cradled here.
Here lived the gentle peasant prince,
    The loving cottar-king,
Compared with whom the greatest lord
    Is but a titled thing.

'Tis but a cot roofed in with straw,
    A hovel made of clay;
One door shuts out the snow and storm,
    One window greets the day;
And yet I stand within this room
    And hold all thrones in scorn,
For here, beneath this lowly thatch,
    Love's sweetest bard was born.

Within this hallowed hut I feel
    Like one who clasps a shrine,
When the glad lips, at last, have touched,
    The something deemed divine!
And here the world, through all the years,
    As long as day returns,
The tribute of its love and tears
    Will pay to ROBERT BURNS!

A Poet King.

Written for the Inauguration of the Glasgow Burns' Statue, which was unveiled by Lord Houghton on the 25th of January, 1877.

JOHN MACFARLANE.

WHAT meaneth this wild commotion?
  Why surgeth the crowd along?
'Tis the natal day of a poet king,
  The chief of Scottish song;
And lo! they come in thousands
  From mountain and strath and glen,
As free in soul as the air they breathe,
  To honour a Saul of men.

And grandly, hark! is ringing
  On the silv'ry streams of day,
"The rank is but of the coin the stamp,
  The man's the gold for aye."
No lyric dream is this,
  To thrill with its magic thrall,
No fancy caught from the wilds of thought,
  But a cry from the hearts of all.

The soul of manhood leaps
  In the toil-encircled throng,

A POET KING.

They shake the earth with their bounding tread,
　For he hath made them strong ;
For wreathed with the light of genius,
　The labour-warrior stands,
And the bulwarks e'en of a throne might fall
　If smote by his horny hands.

And the blinded god of Mammon,
　Hath paled at the minstrel's name,
And a shiver hath passed to his crusted soul
　'Neath the blaze of the heavenly flame ;
The tyrant with gloom in his heart,
　And the brand of Cain on his brow,
Like a craven quakes in his white-lipp'd fear,
　At the gleaming of Freedom now.

　　．　　．　　．　　．　　．　　．

The shroud of the past hath vanished,
　And the mighty-given-of-God,
Looms forth entranced with the meanest flower,
　That springs from the verdant sod ;
Oh ! wildly impassioned spirit !
　In the throes of thy great unrest,
Thou gavest the golden chalice of Thought,
　But we called for the ribald jest.

The stamp of the mind unfettered,
　The smile and the orbèd fire,
No magic touch to the image brings,
　We garnish a broken lyre :
But scarr'd with the fight of ages,
　Triumphantly Scotia turns,
With a queenly glance of pride in her eyes,
　To gaze on her laureate BURNS.

Rantin' Robin.
A Song for Burns's Anniversary.

A. H. WINGFIELD.

WE'VE met this nicht in honour o' Auld Scotia's peasant bard,
Tho' mony years has past since he has gone to his reward,
Ilk twenty-fifth o' Jan'wary we roose the " thackit cot,"
An' drink to ROBIN'S mem'ry in a wee drappie o't.

The saintly cynic whiles may sneer at ROBERT BURNS'S name,
The cantin' hypocrite may jeer, an' cry out "fie for shame,"
But let them jeer, an' let them sneer, there's no ae honest Scot
But pledges Rantin' ROBIN in a wee drappie o't.

His " lilts upon the doric lyre" has pleased baith great an' sma',
His "Cottar's Saturday at e'en " has often charmed us a';
The "Daisy" crush'd amang the stoor, the "thistle's" jaggie coat,
An' "peck o' maut" frae whilk he preed the wee drappie o't.

'Twas he wha said—wi' reason for't—that " man was made to mourn,"
'Twas he wha sang sae nobly o' the Bruce o' Bannockburn,

'Twas he wha proved "a man's a man," tho' poortith be his lot,
If honest, tho' he whiles may tak' a wee drappie o't.

O sweet he sang o' "Bonnie Doon" an' witchin' "Hallowe'en,"
O "Corn rigs, an' shorn rigs," o' "Mary" an' o' Jean ";
The "Limpin' hare," the "Haggis" rare an' "Jenny's' luckless lot,
The reamin' horn on New-Year's morn, an' wee drappie o't.

"Should auld acquaintance be forgot?" ah no, we'll never tine,
Our love for BURNS it's woven in our hearts wi' "Auld Langsyne."
The ae best fellow e'er was born, the independent Scot,
Wha sang auld Scotia's howes an' knowes, an' wee drappie o't.

On "a' the airts the win' can blaw" will ROBIN's fame be borne,
In spite o' they who try to haud his mem'ry up to scorn,
While surges roar o'er Berwick-law he ne'er will be forgot,
Or Tintock's cap contains a drap—a wee drappie o't.

To the Memory of Burns.

Read at a Meeting held in Commemoration of the Poet's Birth.

FRANCIS BENNOCH.

IMMORTAL Bard !—immortal BURNS !
The patriot and the prince of song,
When friends are met shall they forget
The honours which to thee belong ?
    Immortal BURNS !

In every land where truth is known,
The music of thy god-like mind
In strains of melting love hath flown
To fraternize the human kind
    Immortal BURNS !

Thy lays have sear'd the tyrant's heart,
Like flaming bars of hottest steel,
But rais'd the poor to know their right
To think as men—as men to feel,—
    Immortal BURNS !

When light and hope, and reason die,
And darknes shrouds the face of day,

And all things fade,—O, only then
 Shall Scotland's Bard in fame decay.
  Immortal BURNS!

With reverent silence we will fill
 A cup when'er this day returns,
And pledge the memory of the Bard,
 The Bard of Nature—ROBERT BURNS.
  Immortal BURNS!

Address to Burns.

JAMES D. CRICHTON.

The circling wheels of Time have roll'd,
   And brings the fatal day again,
When Death's dark wings swoop'd to enfold
   Thy spirit—king of men !
No idle pomp thy kinghood mock'd—
   A peasant father's hope and joy ;
A peasant dame the cradle rock'd
   That held her black-eyed boy.

Storm mark'd thy entrance into life,
   Wild blew the blast that Januar' morn,
Symbolic of turmoil and strife
   To which the babe was born.
Grim poverty and sad-eyed care,
   Twin sisters, stood beside thy cot ;
They look'd on their unconscious heir,
   And dowered him with their lot.

It was not in baronial hall
   Or mansion proud that thou was't bred,
A cottar's shieling poor and small
   Sheltered thy infant head.
But there were virtues 'neath that roof
   Gracing but rarely loftier rank,
There evil met with stern reproof,
   Good, recognition frank.

There when the toilsome day was sped
  The family cluster'd round the board,
The holy book by turns they read
  With Wisdom's teachings stor'd ;
And when the evening psalm was sung
  The peasant father, old and gray,
Surrounded by his children young
  Knelt humbly down to pray.

The lessons that he taught thee then
  Were not forgot in manhood's prime,
Though happily he might not ken
  Grave errors of that time;
But when that father was no more,
  Thou shouldering brave the heavy load,
Toiled to augment the children's store,
  And with them worshipp'd God !

Hard task to cleave the stubborn soil
  That barely might a pittance yield,
And make a golden harvest smile
  Above a barren field.
Lochlea thy patient struggles saw,
  Tasking thy strength both late and air,
And Mossgiel's wind-swept upland raw
  Thy hour of dark despair !

Yet there were raptures known to thee
  When friends were powerless to condole,
That cheer'd thy breast in penury
  And charm'd thy poet-soul,
For while thy ploughshare turn'd the sod
  The hovering Muse benignly smil'd,
And with the best gift sent from God
  Dower'd her wayward child.

'Twas *her* word opened that inner eye,
  And visions fill'd thy teeming brain,
Thrilling thy soul with ecstacy
  Almost akin to pain.
'Twas *her* touch loos'd thy tongue to sing,
  While grandeur look'd askance with scorn,
Till all who heard its accents ring
  Hail'd a new poet born!

Well didst thou love fair Nature's face
  (For everything to thee was good),
And in thy verse rejoice to trace
  Her every changing mood.
Nothing was common in thy sight,
  And nought escap'd thy eagle ken
That rang'd from red-tipp'd daisies bright
  To dogs and mice and men.

Hast thou not painted for all time
  The lovely scenes where thou didst dwell,
Till natives of a tropic clime
  Know BURNS' haunts so well—
The "banks and braes" of winding Doon,
  Whose waters glass the drooping birk,
Auld Ayr's "twa brigs" and ancient "toon,"
  And Alloway's haunted kirk?

How warmly glow'd the sacred fire
  Of patriot fervour in thy breast,
Dear Scotland's praises to inspire
  And give them added zest.
Lo! Freedom there her altar rear'd—
  Where could she find a nobler shrine?
And lit her torch by despots fear'd—
  At that pure flame divine!

Deep is the debt we owe of praise
  To thee, who with enchanter's wand
Reviv'd in purer form the lays
  Of thy dear native land.
When Scotia takes her place beside
  The cultur'd nations of to-day,
She points to thee, with grateful pride,
  The Bard of "Scots wha hae!"

O sorcerer of mighty skill
  What glamour was it thine to cast,
To waken love's delirious thrill,
  And hold it firm and fast?
For while thy genius kept spell-bound
  The sober, philosophic mind,
Thy lighter hours with love were crown'd,
  And woman aye was kind.

As butterfly from flower to flower,
  So thou from love to love didst flit,
Each fancy had its passing hour,
  Each hour its fever-fit.
Yet amid recklessness and sin,
  How much surviv'd of pure and good,
That promis'd higher meed to win
  By nobler paths pursued?

O bright was that brief day of love
  When Highland Mary cross'd thy way,
And wander'd with thee through the grove
  The "lee-lang simmer day."
True was the mutual love ye swore,
  Dipping your hands into the stream :
Raptures, alas! too quickly o'er,
  Sad waking from thy dream!

But first and last and nobly best,
　Of all the flames thy bosom knew,
Was Jean's affection pure and blest,
　Thy wife, so meek and true.
Constant and calm its ray would burn
　While thou wert heedless straying far,
She knew thy wandering heart would turn
　Back to its guiding star !

'Twas "thoughtless follies" laid thee low,
　Thine own apology is best ;
And we who read thy verse may know
　These follies self-confess'd.
Then let us gently o'er them pass,
　For who of erring mortals can
The first stone cast at one who was
　Not more, nor less, than man?

Thou need'st not fear thy fame will die !
　While blood in Scottish veins runs hot
Thy verse we'll hold in memory,
　Thy faults alone forgot.
For nobler name was ne'er inscrib'd
　On Roman cinerary urns
Of king or bard—a nation's pride—
　Than thy name—ROBERT BURNS !

The century has well nigh flown,
　And finds us honouring now as then,
And thou hast only dearer grown
　To thy fond countrymen.
Still king of men and bards thou art,
　O BURNS ! thou needst not fear eclipse,
True poesy rooted in thy heart
　And blossom'd on thy lips !

To the Memory of Robert Burns.

EDWARD RUSHTON.

Poor, wildly sweet uncultur'd flower,
Thou lowliest of the Muse's bow'r,
"Stern ruin's ploughshare, 'mang the stowre,
    Has crushed thy stem,"
And sorrowing verse shall mark the hour,
    "Thou bonnie gem."

'Neath the green turf, dear Nature's child,
Sublime, pathetic, artless, wild,
Of all thy quips and cranks despoil'd,
    Cold dost thou lie;
And many a youth and maiden mild
    Shall o'er thee sigh.

Those pow'rs that eagle-wing'd could soar,
That heart which ne'er was cold before,
That tongue which caus'd the table roar,
    Are now laid low,
And Scotia's sons shall hear no more
    Thy rapt'rous flow.

Warm'd with "a spark o' Nature's fire,"
From the rough plough thou didst aspire
To make a sordid world admire;
    And few like thee,
Oh, BURNS! have swept the minstrel's lyre
    With ecstacy.

TO THE MEMORY OF BURNS.

Ere winter's icy vapours fail,
The violet in the uncultur'd dale,
So sweetly scents the passing gale,
    That shepherd boys,
Led by the fragrance they inhale,
    Soon find their prize.

So when to life's chill glens confin'd,
Thy rich, tho' rough untutor'd mind,
Pour'd on the sense of each rude hind
    Such sonsie lays,
That to thy brow was soon assign'd
    The wreath of praise.

Anon, with nobler daring blest,
The wild notes throbbing in thy breast,
Of friends, wealth, learning unpossess'd,
    Thy fervid mind
Tow'rds Fame's proud turrets boldly press'd,
    And pleased mankind.

But what avail'd thy pow'rs to please,
When want approach'd, and pale disease ;
Could these thy infant brood appease
    That wail'd for bread ?
Or could they, for a moment, ease
    Thy woe-worn head ?

Applause, poor child of minstrelsy,
Was all the world e'er gave to thee ;
Unmov'd, by pinching penury
    They saw thee torn,
And now, kind souls ! with sympathy,
    Thy loss they mourn.

TO THE MEMORY OF BURNS.

Oh! how I loathe the bloated train,
Who oft had heard thy dulcet strain;
Yet, when thy frame was rack'd with pain,
    Could keep aloof,
And eye with opulent disdain
    Thy lowly roof.

Yes, proud Dumfries, oh! would to heaven
Thou hadst from that cold spot been driven,
Thou might'st have found some shelt'ring haven
    On this side Tweed :—
Yet, ah! e'en here, poor bards have striven,
    And died in need.

True genius scorns to flatter knaves,
Or crouch amidst a race of slaves;
His soul, while fierce the tempest raves,
    No tremour knows,
And with unshaken nerve he braves
    Life's pelting woes.

No wonder, then, that thou should'st find
Th' averted glance of half mankind;
Should'st see the sly, slow, supple mind
    To wealth aspire,
While scorn, neglect, and want combin'd
    To quench thy fire.

While wintry winds pipe loud and strong,
The high-perch'd storm-cock pours his song;
So thy Æolian lyre was strung
    'Midst chilling times;
Yet clearly didst thou roll along
    Thy "routh of rhymes."

And oh ! that routh of rhymes shall raise
For thee a lasting pile of praise.
Haply some wing, in these our days,
  Has loftier soar'd :
But from the heart more melting lays
  Were never pour'd.

Where Ganges rolls his yellow tide,
Where blest Columbus' waters glide,
Old Scotia's sons, spread far and wide,
  Shall oft rehearse,
With sorrow some, but all with pride,
  Thy 'witching verse.

In early spring, thy earthly bed
Shall be with many a wild flow'r spread ;
The violet there her sweets shall shed,
  In humble guise,
And there the mountain-daisy's head
  Shall duly rise.

While darkness reigns, should bigotry,
With boiling blood, and bended knee,
Scatter the weeds of infamy
  O'er thy cold clay,
Those weeds, at light's first blush, shall be
  Soon swept away.

And when thy scorners are no more,
The lonely glens, and sea-beat shore,
Where thou hast croon'd thy fancies o'er
  With soul elate,
Oft shall the bard at eve explore,
  And mourn thy fate.

Robert Burns.

(Written for the Centennial Celebration of 1859.)

EVAN MACCOLL.

So many minstrels known to fame
Have made sweet Coila's bard their theme,
That like an oft-told tale may seem
    All *I* can sing of ROBIN.
Yet be his cairn however high,
No Scot can mutely pass it by;
The tribute of a song and sigh
    Let's therefore give to ROBIN.

His was the true poetic art
To sing directly from the heart:
To waken mirth, or tears to start,
    No mortal matches ROBIN!
Now gently flow his thoughts along,
Now, like a rushing river strong,
A very cataract of song
    Resistless is our ROBIN!

The sun not aye unclouded shines;
There's dross within earth's richest mines;—
ROB had his faults, and grave divines
    Oft shook their heads at ROBIN.

A lassie "coming through the rye"
Unkissed, he never could pass by;
Nor can I blame him much, for why,
    The lasses all loved ROBIN!

ROB loved to speak the truth right down,
No matter who might smile or frown;
A rascal, be he king or clown,
    No mercy had from ROBIN.
His sympathies—how dread to tell!
Embraced all being—Nick himsel'—
Yes, pity for the very de'il,
    No sin or shame thought ROBIN.

I see him with scorn-flashing eyes
Detect "a cuif" in lordly guise;
To see was to denounce—despise:
    "A man's a man," quoth ROBIN!
Hold, honest Labour, up thy head,
And point with pride to ROBIN dead;
The halo round thy path he shed
    Immortal is as ROBIN.

Alas, that not till they are lost
The gifts that we should value most
Are rightly prized! To Scotland's cost,
    Thus fared it with her ROBIN.
Yet may she glory loud and long
To know, of all earth's sons of song,
The most world-honoured of the throng
    Is Coila's matchless ROBIN!

On the Death of Burns.

Mrs. Grant of Laggan.

WHAT adverse fate awaits the tuneful train?
Has Otway died and Spencer liv'd in vain?
In vain has Collins, Fancy's pensive child,
Pour'd his lone plaints by Avon's windings wild?
And Savage, on Misfortune's bosom bred,
Bared to the howling storm his houseless head?
Who gentle Shenstone's fate can hear unmoved,
By virtue, elegance, and genius lov'd?
Yet, pensive wand'ring o'er his native plain?
His plaints confess'd he lov'd the Muse in vain.
Chill penury invades his favourite bower,
Blasts every scene, and withers every flower;
His warning Muse to Prudence turn'd her strain,
But Prudence sings to thoughtless bards in vain;
Still restless fancy drives them headlong on
With dreams of wealth, and friends, and laurels won—
On ruin's brink they sleep, and wake undone!
And see where Caledonia's genius mourns,
And plants the holly round the grave of BURNS!
But late "its polished leaves and berries red
Play'd graceful round the rural Poet's head;"
And while with manly force and native fire
He wak'd the genuine Caledonian lyre,
Tweed's severing flood exulting heard her tell,
Not Roman wreaths the holly could excel;
Not Tiber's stream, along Campania's plain,
More pleas'd, convey'd the gay Horatian strain,
Than bonnie Doon, or fairy-haunted Ayr,
That wont his rustic melody to share,

Resound along their banks the pleasing theme,
Sweet as their murmurs, copious as their stream :
And Ramsay, once the Horace of the North,
Who charm'd with varied strains the listening Forth,
Bequeath'd to him the shrewd peculiar art
To satire nameless graces to impart,
To wield her weapons with such sportive ease,
That, while they wound, they dazzle and they please :
But when he sung to the attentive plain
The humble virtues of the patriarch swain,
His evening worship, and his social meal,
And all a parent's pious heart can feel ;
To genuine worth we bow submissive down,
And wish the Cottar's lowly shed our own :
With fond regard our native land we view,
Its cluster'd hamlets, and its mountains blue,
Our " virtuous populace," a nobler boast
Than all the wealth of either India's coast.
Yet while our hearts with admiration burn,
Too soon we learn that " man was made to mourn."
The independent wish, the taste refin'd,
Bright energies of the superior mind,
And feeling's generous pangs, and fancy's glow,
And all that liberal nature could bestow,
To him profusely given, yet given in vain ;
Misfortune aids and points the stings of pain.
How blest when wand'ring by his native Ayr,
He woo'd "the willing Muse," unknown to care !.
But when fond admiration spread his name,
A candidate for fortune and for fame,
In evil hour he left the tranquil shade
Where youth and love with hope and fancy play'd ;
Yet rainbow colours gild the novel scene,
Deceitful fortune sweetly smil'd like Jean ;

ON THE DEATH OF BURNS.

Now courted oft by the licentious gay,
With them through devious paths behold him stray;
The opening rose conceals the latent thorn,
Convivial hours prolong'd awake the morn,
Even reason's sacred pow'r is drown'd in wine,
And genius lays her wreath on folly's shrine;
Too sure, alas! the world's unfeeling train
Corrupt the simple manners of the swain;
The blushing muse indignant scorns his lays,
And fortune frowns, and honest fame decays,
Till low on earth he lays his sorrowing head,
And sinks untimely 'midst the vulgar dead!
Yet while for him, belov'd, admir'd in vain,
Thus fond regret pours forth her plaintive strain,
While fancy, feeling, taste, their griefs rehearse,
And deck with artless tears his mournful hearse,
See cunning, dulness, ignorance, and pride,
Exulting o'er his grave in triumph ride,
And boast, "tho' genius, humour, wit, agree,"
Cold selfish prudence far excels the three;
Nor think, while groveling on the earth they go,
How few can mount so high to fall so low.
Thus Vandals, Goths, and Huns, exulting come,
T' insult the ruins of majestic Rome.
But ye who honour genius—sacred beam!
From holy light a bright etherial gleam,
Ye whom his happier verse has taught to glow,
Now to his ashes pay the debt you owe,
Draw pity's veil o'er his concluding scene,
And let the stream of bounty flow for Jean!
The mourning matron and her infant train
Will own you did not love the muse in vain,
While sympathy with liberal hand appears,
To aid the orphan's wants, and dry the widow's tears!

Stanzas

Written on a Copy of the Engraving of Robert Burns from Taylor's lately recovered Picture.

THOMAS ATKINSON.

AND this was Scotland's noblest son of song !
  How calm his mien—how sadly still his look !
Where be the flashes, bright and brief, yet strong,
  Of mirth that revels, though the wise rebuke?
  Tell me, thou limner, in what sacred nook
Of this expanse of chasten'd countenance,
  There lurk'd the gibe and jest which often shook
The stolid crowd—in wit's omnipotence?
Why live not these in this—and where their recompense?

Lurks the rich treasure in that placid gaze—
  In the deep meaning of those full-orb'd eyes—
In the veil'd lustre which, as through a haze
  Of mellowing beauty, meekly lifts the guise
  Of mere humanity, and shows what lies
In the far chambers of the soul still kept?
  It does !—it does !—and O ! more dear I prize
The soft, yet manly sadness that hath crept
O'er this, than would I all the heights by art o'erleapt.

STANZAS.

Look ! what a brow soars o'er these arched spells,
   That fix my gaze, they look so sad on me !
See ! where hid meaning into language swells
   Upon these lips, that seem as tremblingly
   To heave, as leaves upon a wind-woo'd tree !
Yet prophet power hath touch'd them with its fire ;
   With burning balm love dew'd them thrillingly !
Have they not blazed, like lightning on a pyre,
As from them flash'd the words that speak a patriot's ire ?

O ! it is deeply true—no transient glance
   Can tell the meaning of the Poet's look ;
For who will say, who on one mood may chance
   To wondering gaze, that he hath not mistook
   The hue the moment's inspiration took ;
For the deep shadows that from others hide
   The broken hopes, the soul's self-urged rebuke,
Which in his breast for ever might abide,
Converting into gall, dear BURNS ! thy heart's warm tide !

Written for Burns' Anniversary.

ROBERT ALLAN.

WHEN Januar's winds sae fiercely blaw,
An' drive alang the drifting snaw,
It's roun' the ingle then we ca'
    The merry tales o' ROBIN.
We vow he was a man o' worth,
The pride an' honour o' the north;
An' though he's cauld now i' the earth
    We think aye weel o' ROBIN.

We canna turn a page or twa,
But on a line or verse we'll fa'
That's dear to Caledonia,
    An' worthy aye o' ROBIN.
His vera name, it is a charm
That a' our hearts at ance can warm :
The deil be on them that wad harm
    The memory o' ROBIN.

The ploughman whistling at his plough,
The mountain daisy wat wi' dew,
The blythe birds sporting on the bough,
    Inspired the heart o' ROBIN.

Fond lovers 'neath the milk-white thorn,
The farmer by his waving corn,
The dewy eve, the dawning morn,
    Aye cheered the heart o' ROBIN.

Sae ready wi' his jokes an' rhymes,
Lord help them that were read in crimes !
The vera priests themselves betimes
    Wad stan' in awe o' ROBIN :
An' mair for token, let me tell,
He didna spare the deil himsel',
But tauld him a' his fauts pell-mell,—
    He ne'er met ane like ROBIN.

What he has dune in prose an' verse
We are na fit here to rehearse ;
Sam Johnson wad o' words be scarce
    To sing the praise o' ROBIN.
O' poets Scotland has her share,
An' some o' pith and spirit rare,
But where's the ane that can compare
    Wi' our immortal ROBIN?

Let monuments, hy men o' art,
An' pillars up like mushrooms start,
There's nae mausoleum like the heart
    That thinks aye weel o' ROBIN :
Sae let us a' in merry tune,
Wi' hearts life's ills an' cares aboon,
Here drink ance mair, as aft we've dune,
    The memory o' ROBIN.

Thoughts

On Visiting the Grave of Burns.

ALEXANDER MACLAGGAN.

THE loud voice of a stormy e'en
Came raving to our cottage pane ;
The cottar bodies closed their een
    In sleep, to shun
Dreigh sights, that they a' day had seen
    Deface the sun.

Unmindfu' o' the raging blast,—
Though heaven to earth was fa'in fast,—
O'er hill, an' heath, an' field I pass'd
    By eerie turns,
To view the dark—the lone—the last
    Abode of BURNS.

The grave of BURNS ! a throne of state !
Revered, though mouldering desolate !
I cursed fell poortith's hapless fate
    And quick decay,
As musing on the "furrow's weight"
    That o'er him lay.

His morn of life in darkness rose,
But darker still its dreary close;
I' the space between, unnumber'd woes
   Were on him hurled;
Yet, from his darkness, light arose
   That glads the world.

O, matchless BURNS ! that I'd been livin'
When the power o' sang to thee was given,
And seen, when misery had riven
   Thy manly form,
Thy soul, the undying gift of Heaven,
   Defy the storm !

Or seen thee in a calmer hour,
When o'er thee bent the blooming bower;
Or gazing on the crimson flower,
   The daisy fair,
And heard thee bless the Almighty power
   Who placed it there :

Or seen thee in a lonely shade,
Fast wrapping in the rustic plaid
Thy Mary—dear departed maid !
   In fond embrace,
And mark'd the game fond passion played
   Upon thy face :

Or seen thee in thy hours o' glee,
Wild, bold, and witty, frank and free,
Keen joining on the flowery lea
   The rustic dance,
And watchin' frae Jean's lowing e'e
   Love's kindling glance !

Or seen thee by the ingle-nook,
When wi' thy jest the biggin' shook;
Or stalkin' by the oaten stook,
    Frae man afar,
When heavenward went thy passionate look
    To the " lingering star."

Many are they who would aspire
To wake again thy sleeping lyre,
Wasting their breath to blow a fire,
    To burn like thine;
But black I see them all expire
    Before thy shrine!

BURNS! might I live again to see
A bard among us like to thee,
My heart's best thanks I glad would gie
    To God, the giver—
Then in contentment close my e'e,
    To sleep for ever.

Song

For the Anniversary of the Birth-Day of Robert Burns.

ANDREW PARK.

BRAVE Scotland—Freedom's throne on earth !—
   A bumper to thy glory ;
This day thy matchless Bard had birth,
   So famed in song and story.
Where'er thy mountain-sons may stray,
   Thou'st thrown thy magic round them,
And on this ever-hallow'd day,
   In kindred love has bound them.

He nobly walk'd behind his plough,
   And gazed entranced on Nature ;
While genius graced his lofty brow,
   And play'd in every feature !
For then inspired by glowing songs
   Of " Bruce,"—or " Highland Mary,"
The minstrel-birds, in joyous throngs,
   Around their Bard would tarry !

But wae's my heart—he sings nae mair
   In strains o' joy or sorrow,
Though on the bonnie banks o' Ayr.
   His spirit smiles each morrow !

SONG.

Aud Scotia's name—enthroned on high—
  The great, the gentle hearted!
Sits with the tear-drop in her eye,
  And mourns her Bard departed!

O sacred land of gallant men!—
  Of maiden unassuming!
Who dwell obscure by loch and glen,
  Where still the thistle's blooming!
How well has BURNS rehearsed your praise—
  Among your cloud-crown'd mountains,
In never-dying, tuneful lays,—
  Pure as your native fountains!

Then fill the sparkling goblet high,
  And let no discord stain it;—
Let joy illume each manly eye,
  While to the dregs we drain it!
To BURNS! to BURNS!—the king of song!—
  Whose lyre shall charm all ages;
Mirth, wisdom, love, and satire strong
  Adorn his deathless pages.

Robert Burns.

Joseph Cunningham.

Hail, Caledonia ! land of song and story,—
    Land of the fair, the virtuous and the brave !
The brightest star that sheds on thee its glory
    Rose from the darkness of thy Burns's grave :
That star shall be a light among the nations
    When prouder orbs have faded and grown dim,
And hailed with pride by coming generations,
    For man yet knows not all he owes to him.

His strains have nerved the feeble 'gainst oppression,
    Aroused in true men's hearts a scorn of wrong,—
Pointed the hopeless to man's sure progression,
    And taught the weak to suffer and be strong.
Lessons like these the soul of man shall cherish,
    While through his heart the ardent life-blood springs :
One burning thought, at least, can never perish—
    An honest man's above the might of kings.

While noble souls shall glow with warm emotion,
    While woman loves and genius pants for fame,—
While truth and freedom claim man's deep devotion,
    True hearts shall throb responsive to his name.
Then weep not, Scotland, though thy minstrel slumbers ;
    Still lives the spirit of his song sublime,—
Still shall the music of his deathless numbers
    Thrill in all hearts and vibrate through all time.

Birth-Place of Robert Burns.

THOMAS WILLIAM PARSONS.

A LOWLY roof of simple thatch—
  No home of pride, of pomp, and sin—
So freely let us lift the latch,
  The willing latch that says " come in."

Plain dwelling this ! a narrow door—
  No carpet by soft sandals trod,
But just for peasant's feet a floor,—
  Small kingdom for a child of God !

Yet here was Scotland's noblest born,
  And here Apollo chose to light ;
And here those large eyes hailed the morn
  That had for beauty such a sight !

There, as the glorious infant lay,
  Some angel fanned him with his wing.
And whispered, " Dawn upon the day
  Like a new sun ! go forth and sing !"

He rose and sang, and Scotland heard—
 The round world echoed with his song,
And hearts in every land were stirred
 With love, and joy, and scorn of wrong.

Some their cold lips disdainful curled;
 Yet the sweet lays would many learn;
But he went singing through the world,
 In most melodious unconcern.

For flowers will grow, and showers will fall,
 And clouds will travel o'er the sky;
And the great God, who cares for all,
 He will not let his darlings die.

But they shall sing in spite of men,
 In spite of poverty and shame,
And show the world the poet's pen
 May match the sword in winning fame.

To the Memory of Robert Burns.

JAMES MACFARLAN.

In lonely hut and lordly hall a mighty voice is heard,
And 'neath its wild bewitching spell the honest brows are bared ;
From Scotland's hills and twilight glens, to far Columbian floods,
It stirs the city's streets of toil, and wakes its solitudes ;
It speaks no triumph reap'd with swords, it brings no conquering cry
Of buried honours, battle-crown'd, and veil'd with victory ;
But hearts leap loving to its note, and kindling bosoms glow,
To hail the Poet born to fame, a hundred years ago.

O, like a glorious bird of God, he leapt up from the earth !
A lark in song's exalted heaven, a robin by the hearth ;
O, like a peerless flower he sprang from Nature's meanest sod,
Yet shedding joy on every path by human footstep trod.
How shall we tell his wondrous power, how shall we say or sing
What magic to a million hearts his deathless strains can bring ?

How men on murkest battle-fields have felt the potent charm,
Till sinking valour leapt to life, and strung the nerveless arm?

How hearts in dreariest loneliness have toil'd through barren brine—
The only glimpse of sunshine then, *his* pictures o' langsyne;
How far amid the western wilds, by one enchanting tune,
The wide Missouri fades away in dreams of "bonny Doon."
More hearts and hands renew the pledge—sweet pledge of other years—
That sacred "auld acquaintance" vow, the light of parting tears.

O! blessed be the brawny arm that tore presumption down,
That snatch'd the robe from worthless pride, and gave to toil a crown;
That smote the rock of poverty with song's enchanting rod·
Till joy into a million hearts in streams of beauty flow'd;
And while that arm could stretch to heaven, and wield the lightning's dart,
It brought the glorious sunshine, too, to cheer the humblest heart:
For free as Spring, his gladsome muse danc'd o'er the daisied plain,
Or rang in organ-gusts of praise through grandeur's mightiest fane.
Then blest for ever be the soul that link'd us man to man—
A brotherhood of beating hearts—God's own immortal plan:
While Labour, smiling at his forge, or stalking at his plough,
Looks up with prouder soul to find God's finger on his brow—

Feels man is man, though russet-robed, and smacking of
   the soil,
And all are brothers, whether born to titles or to toil.

Then pledge his mem'ry far and near, although the hand
   be dust
That oft has swept the golden lyre, that ages cannot rust :
No sun of time e'er sits upon the empire of his fame,
And still unwearied is the wing that bears abroad his name.
There may be grander bards than he, there may be loftier
   songs,
But none have touch'd with nobler nerve the poor man's
   rights and wrongs :
Then, while unto the hazy past the eye of fancy turns,
Raise high the fame, and bless the name of glorious ROBERT
   BURNS.

Ye may talk o' your Learning.

Andrew Mercer.

Ye may talk o' your learning, and talk o' your schools,
  An' how they mak' boobies sae clever;
Gude sooth? ye will never mak' wise men o' fools,
  Altho' ye should study for ever.
If poor be the soil, ye may labour an' toil
  On a common where naething will grow man;
But 'gainst sic barren sods I will lay you some odds
  On the head of an Ayrshire ploughman.

Book-lear' an' the like o't, an' a' the fine things
  That ye hear an' ye get at the college,
If there's no something *here* that school-craft quite dings,
  At best ye're a botch-potch o' knowledge.
But ye've heard o' a heckler wha wound i' the west,
  To whom Nature had gi'en sic a pow, man,
The brairds o' his brain excell'd ither folks' best,
  An' mony ran after his tow, man.

What signifies polish without there be pith?
  Mind that, a' ye gets o' Apollo;
A farmer ance dwelt by the banks o' the Nith,
  By my sang, he wad beat you a' hollow;

For he sang, an' he sowed, an' he penned an' he ploughed,
   An though his barnyard was but sorry,
Frae his girnal o' brain he sowed siccan grain,
   As produced him a harvest of glory.

Ance mair, a poor fallow there dwelt in the south,
   An' he to his trade was a gauger—
He excelled a' the songsters, the auld an' the young,
   I'll haud you a pint for a wager.
I farther might tell, he'd a mind like a stell,
   An' such was his wonderfu' merits,
That the haill country rang, an' the haill country sang,
   When they tasted the strength o' his spirits.

Now wha was this ploughman and heckler sae braw,
   An' wha was this farmer-exciseman?
It was just ROBIN BURNS—for he was them a'—
   An' ye ken that I dinna tell lies, man.
So here's to his memory again an' again,
   Tho' learning is guid, we ne'er doubt it,
But a bumper to him wha had got sic a brain
   That could do just as weel maist without it!

The night you quoted Burns to me.

JAMES NEWTON MATTHEWS.

THE winds of early autumn blew
   Across the midnight. Overhead
   A wild moon up the heavens fled,
And cut the sable vault in two;
We heard the river lap and flow,
   We turned our poet fancies free;
My heart did all its cares forego,
   The night you quoted BURNS to me.

A gray owl from a blasted limb
   Dropped down the dark, and blundered by,
   As if a fiend with flaming eye
Fast followed in pursuit of him;
Ah then you crooned beneath the moon,
   A ditty weird as weird could be;
And Tam O'Shanter crossed the Doon,
   The night you quoted BURNS to me.

We praised the "Lass of Ballochmyle,"
   We talked of Mary loved and lost,
   Until our spirits touched and crossed,
And melted into tears the while.

THE NIGHT YOU QUOTED BURNS.

We drank to " Nell " and " Bonny Jean,"
   To Chloris and " The Banks o' Cree,"
Blest hour—I keep its mem'ry green
   The night you quoted BURNS to me.

The Wabash hills their heads low hung,
   As floating up their winding ways
   They caught the sound of " Logan Braes,"
And heard sweet Afton's glory sung;
And loud the Wabash did deplore
   That no brave poet voice had she,
To lend her fame for ever more:
   The night you quoted BURNS to me.

O dear delightful Autumn day
   For ever gone·beyond recall;
   Comrade, the clouds are over all,
And you—you've vanish'd from my sight;
Still flows the river as of yore,
   The owl still haunts the lonely tree,
And I'll forget—ah, never more,
   The night you quoted BURNS to me.

Birth of Burns.

THOMAS MILLER.

UPON a stormy winter night
Scotland's bright star first rose in sight ;
Beaming upon as wild a sky
As ever to prophetic eye
Proclaimed that Nature had on hand
Some work to glorify the land.
Within a lonely cot of clay,
That night her great creation lay.

Coila—the nymph who round his brow
Twined the red-berried holly-bough—
Her swift-winged heralds sent abroad,
To summon to that bleak abode
All who on genius still attend,
For good or evil to the end.

They came obedient to her call :—
The immortal infant knew them all.

Sorrow and Poverty—sad pair—
Came shivering through the wintry air :
Hope, with her calm eyes, fixed on Time,
His crooked scythe hung with flakes of rime :
Fancy, who loves abroad to roam,
Flew gladly to that humble home :

Pity and Love, who, hand in hand,
Did by the sleeping infant stand :
Wit, with a harum-scarum grace,
Who smiled at Laughter's dimpled face :
Labour, who came with sturdy tread,
By high-souled Independence led :
Care, who sat noiseless on the floor ;
While Wealth stood up outside the door :
Looking with scorn on all who came,
Until he heard the voice of Fame,
And then he bowed down to the ground :—
Fame looked on Wealth with eyes profound,
Then passed in without sign or sound.

Then Coila raised her hollied brow,
And said, "Who will this child endow?"
Said Love, " I'll teach him all my lore,
As it was never taught before ;
Its joys and doubts, its hopes and fears,
Smiles, kisses, sighs, delights, and tears."
Said Pity, " It shall be my part
To gift him with a gentle heart."
Said Independence, "Stout and strong
I'll make it to wage war with wrong."
Said Wit, " He shall have mirth and laughter,
Though all the ills of life come after."
"Warbling her native wood-notes wild,"
Fancy but stooped and kissed the child ;
While through her fall of golden hair
Hope looked down with a smile on Care.

Said Labour, " I will give him bread."
"And I a stone when he is dead,"
Said Wealth, while Shame hung down her head.

BIRTH OF BURNS.

"He'll need no monument," said Fame;
"I'll give him an immortal name;
  When obelisks in ruin fall,
  Proud shall it stand above them all;
  The daisy on the mountain side
  Shall ever spread it far and wide;
  Even the road-side thistle down
  Shall blow abroad his high renown."

Said Time, "That name, while I remain,
Shall still increasing honour gain;
  Till the sun sinks to rise no more,
  And my last sand falls on the shore
  Of that still, dark, and unsailed sea,
  Which opens on Eternity."

Time ceas'd : no sound the silence stirr'd,
Save the soft notes as of a bird
Singing a low sweet plaintive song,
Which murmuring Doon seemed to prolong,
As if the mate it fain would find
Had gone and "left a thorn" behind.

Upon the sleeping infant's face
Each changing note could Coila trace.

Then came a ditty, soft and slow,
Of Love, whose locks were white as snow.

The immortal infant heard a sigh,
As if he knew such love must die.

That ceased : then shrieks and sounds of laughter,
That seemed to shake both roof and rafter,
Floated from where Kirk Alloway
Half buried in the darkness lay.

A mingled look of fun and fear
Did on the infant's face appear.

There was a hush : and then uprose
A strain, which had a holy close,
Such as with Cottar's psalm is blended
After the hard week's labour's ended.
And dawning brings the hallowed day.

In sleep the infant seemed to pray.

Then there was heard the martial tread,
As if some new-born Wallace led
Scotland's armed sons in Freedom's cause.

Stern looked the infant in repose.

The clang of warrior's died away,
And then "a star with lessening ray"
Above the clay-built cottage stood ;
While Ayr poured from its rolling flood
A sad heart-rending melody,
Such as Love chants to Memory,
When of departed joys he sings,
Of " golden hours on angel wings "
Departed, to return no more.

Pity's soft tears fell on the floor,
While Hope spake low, and Love looked pale,
And Sorrow closer drew her veil.

Groans seemed to rend the infant's breast,
Till Coila whispered him to rest ;
And then, uprising, thus she spake :
"This child unto myself I take :

All hail! my own inspired Bard,
In me thy native Muse regard!"
Around the sleeping infant's head
Bright trails of golden glory spread.
"A love of right, a scorn of wrong,"
She said, "unto him shall belong;
A pitying eye for gentle woman,
Knowing 'to step aside is human';
While love in his great heart shall be
A loving spring of poetry.
Failings he shall have, such as all
Were doomed to have at Adam's fall;
But there shall spring above each vice
Some golden flower of Paradise,
Which shall, with its immortal glow,
Half hide the weeds that spread below;
So much of good, so little guile,
As shall make angels weep and smile,
To think how like him they might be
If clothed in frail humanity;
His mirth so close allied to tears,
That when grief saddens or joy cheers,
Like shower and shine in April weather,
The tears and smiles shall meet together.
A child-like heart, a god-like mind,
Simplicity round Genius twined:
So much like other men appear,
That, when he's run his wild career,
The world shall look with wide amaze,
To see what lines of glory blaze
Over the chequered course he passed—
Glories that shall forever last.

Of Highland hut and Lowland home,

His songs shall float across the foam,
Where Scotland's music ne'er before
Rang o'er the far-off ocean shore,
To shut off eve from early morn,
They shall be carolled 'mid the corn,
While maidens hang their heads aside,
Of Hope that lived, and Love that died;
And huntsmen on the mountain's steep,
And herdsmen in the valleys deep,
And virgin's spinning by the fire,
Shall catch some fragment of his lyre.
And the whole land shall all year long
Ring back the echoes of his song,
The world shall in its choice records
Store up his common acts and words,
To be through future ages spread;
And how he looked, and what he said,
Shall in wild wonderment be read,
When coming centuries are dead."

"And wear thou this," she solemn said,
"And bound the holly round his head;
The polished leaves, and berries red,
  Did rustling play;
And like a passing thought she fled
  In light away."

An Evening with Burns.

Agnes Maule Machar.

WITHOUT, the "blast o' Janwar wind"
   About the building seemed to linger,
That on a wintry night "langsyne"
   "Blew hansel in" on Scotland's singer.

Within, we listened, soul attent,
   To tones attuned by tenderest feeling;
The music of the Poet's soul
   Seemed o'er our pulses softly stealing.

We saw again the ploughman lad,
   As by the banks of Ayr he wandered,
With burning eyes and eager heart,
   And first on Song and Scotland pondered.

We saw him as from Nature's soul
   His own drew draughts of joy o'erflowing;
The plover's voice, the briar-rose,
   The tiny harebell lightly growing.

The wounded hare that passed him by,
   The timorous mousie's ruined dwelling,
The cattle cowering from the blast,
   The dying sheep her sorrow telling—

All touched the heart that kept so strong
   Its sympathy with humbler being,
And saw in simplest things of life
   The poetry that waits the seeing.

We saw him 'mid the golden grain
   Learning the oldest of romances;
At first his boyish pulses stirred
   " A bonnie lassie's " gentle glances.

We saw the birk and hawthorn shade
   Droop o'er the tiny running river,
Where he and his dear Highland maid
   Spoke their farewell—alas, for ever!

There be the Poet's wish fulfilled
   That summer ever "langest tarry,"
For all who love the singer's song
   Must love his gentle Highland Mary!

Alas, that other things than these
   Were written on the later pages,
That made that tortured soul of his
   A by-word to the after ages.

For many see the damning sins
   They lightly blame on slight acquaintance,
But *not* the agony of grief
   That proved his passionate repentance.

'Twas his to feel the anguish keen
   Of noblest powers to mortal given,
While tyrant passions chained to earth
   The soul that might have soared to heaven.

'Twas his to feel in one poor heart
  Such war of fierce conflicting feeling,
As makes this life of ours too sad
  A mystery for our unsealing—

The longing for the nobler course,
  The doing of the thing abhorrent—
Because the lower impulse rose
  Resistless as a mountain torrent—

Resistless to a human will,
  But not to strength that had been given,
Had he but grasped the anchor true
  Of "correspondence fixed wi' heaven."

Ah, well! he failed. Yet let us look
  Through tears upon our shining brother,
As thankful that we are not called
  To hold the balance for each other!

And never lips than his have pled
  More tenderly and pitifully,
To leave the erring heart with him
  Who made it, and will judge it truly.

Nay more, it is no idle dream
  That we have heard a voice from heaven,
"*Behold this heart hath lovéd much,
  And much to it shall be forgiven!*"

Kossuth at the Grave of Burns.

ALEXANDER G. MURDOCH.

IMMORTAL picture, fixed in memory's light,
  To which proud Freedom's stretch'd forefinger turns—
Kossuth, the champion of Truth and Right,
  Beside the grave of BURNS !

A nation's-hero, flush'd with light and fame,
  Standing apart, and with uncover'd head,
By his grand dust who worshipp'd Bruce's name—
  The patriot-poet dead !

Dead ! nay, not dead ; the poet never dies ;
  He lives, as lives the hero of the land ;
Proud-set in mausoleum'd memories,
  Truth's torchlight in his hand.

Death gives the bard but longer lease of life ;
  The spirit of his song is never dumb ;
Whole peoples march to it, in peace and strife,
  As armies to the drum.

Kossuth and BURNS ! names deathless as the stars !
  Twin-record of a battle waged for man ;
A red light, threatening, like the soul of Mars,
  Oppression's power and ban.

Soldier of God ! had BURNS been of thy day
  He would have grasp'd thy hand with fervent pride ;
Thou later Wallace ! lighting Hungary's way
  Up Freedom's steep hillside.

Like BURNS, thy passion was the good of man,
  The breaking down of Wrong, and false degree ;
Defying, to the death, each curse and ban
  Would hold men's souls in fee.

Oh, glorious deed to face the spears of Gath ;
  Worthy of liberty's sublime acclaims ;
Kossuth and BURNS ! the world's far onward path
  Is lamp'd by these great names ;

Greater than all the Cæsars, and more lov'd,
  Tho' less gilt with the trapping of success ;
Their's were the fire-swept outposts that back shov'd
  Wrong's arm'd and crushing press.

For, did not BURNS, the Scottish patriot bard,
  Like brave Kossuth, the patriot-hero, stand
A pioneer in the advanced vanguard
  For man, and native land ?

Unhappy BURNS, thine was the martyr's part,
  In that thy life was dash'd with wreck and gloom ;
And meet it is that Kossuth's lion heart
  Should beat so near thy tomb.

That tomb to which, though grander tombs there be,
  The Scottish heart in fond remembrance turns ;
For, holds it not our noblest legacy—
  The dust of ROBERT BURNS ?

THE GRAVE OF BURNS.

Oh, had Kossuth fit words as he stands there,
   His noble heart by strong emotion thrill'd,
How would he pour forth, like a prophet's pray'r,
   His soul with worship fill'd.

Fit subject for some master-painter's brush,
   Kossuth, bare brow'd, at ROBERT BURNS'S tomb,
His mighty soul, absorb'd in solemn hush,
   Giving no mean thought room.

Though mute, doubt not his hero-heart beat high
   While standing by the patriot-poet's urn,
Hearing old Scotland shouting to the sky
   The charge at Bannockburn.

Ah! not for him the triumph-burst that comes
   Once to a hero, in some fateful hour;
Huzza'd to victory 'mid the roll of drums,
   As kings are borne to power.

Too oft, alas! high worth, and genius both,
   Have trod life's path with bruis'd and bleeding feet,
Making of Conscience an inviolate oath—
   The God-voiced paraclete.

Let history with a pen of sorrow tell
   How Hungary, amid a world's regrets,
Went down 'neath leagued oppression's shot and shell
   And blood-stain'd bayonets.

Pale Hungary bled; but with a firm right hand
   Kossuth redeem'd her banner from the fight,
And waved it, exiled from his native land,
   In the world's open sight.

THE GRAVE OF BURNS.

And glorious was the aftermath that came,
  As come it will to all who nobly strive;
The torch he lit burn'd with consuming flame,
  That no wrong could survive.

And now his country stands unshackl'd, free;
  The chains that bound her once are cut in twain;
Kossuth's life-work is her proud legacy—
  Brave Hungary lives again!

High-soul'd Kossuth! thy tears alone may tell
  What thoughts were thine at ROBERT BURNS's tomb;
The bard who, lark-like, sang aloft, then fell
  Through clouds of glory gloom.

Not his, howe'er, the doom that comes of death;
  His genius, shrin'd in words of burning song,
Lives fresh to-day upon a whole world's breath,
  And will not stoop to wrong.

The sun goes down; but high above the night,
  In far-set isolation, spaced alone,
For ever and for ever, orb'd in light,
  The stars of God burn on!

Immortal picture, rescued from the years,
  To which the eye of fancy oft returns;
Kossuth, a century's hero, shedding tears
  Beside the grave of BURNS.

The Burns Monument, Kilmarnock.

Alexander G. Murdoch.

> Go, read the names that know not death,
>   Few nobler ones than BURNS are there;
> And none hath won a grander wreath
>   Than that which binds his hair.—*Halleck.*

I.

I HANDLE life's kaleidoscope, and lo! as round it turns,
I see, beneath an arc of hope, the young boy-poet, BURNS.
Dream-vision'd, all the long rich day he toils with pulse of pride,
Among the sun-gilt ricks of hay, his "Nelly" by his side;
The world to him seems wondrous fair; sunrise and sunset fill
With music all the love-touch'd air, entoned in bird and rill;
Times move apace: the ardent boy confronts life's deep'ning fight,
And "Handsome Nell"—a first-love joy—melts into memory's light.

II.

I look again, and shining noon still finds him chain'd to toil,
His soul-thron'd with the lark song-pois'd above the daisied soil;

Mossgiel! upon thy green sward now the song-king grandly
    stands,
God's sunshine on his face and brow, the plough-horns in
    his hands.
Mouse! that dost run with "bickering gait," stay, stay thy
    trembling flight,
The bard who wept the "Daisy's" fate laments thy hap-
    less plight;
And, perchance, when the gloamin' lies on glen and hill-
    side green,
Thy mishap may re-wet his eyes, told o'er to "Bonnie
    Jean."

III.

Kilmarnock! oft thy streets and lanes echoed the poet's
    tread:
He brought to his matchless strains, asking for fame—not
    bread;
And see, the proud bard, dream-rapt, stands for one sweet
    hour apart,
His book of song within his hands, and in his book—his
    heart!
O, happy town, that gave the bard a gift hope-eloquent,
His dearest wish and first reward—his book in "guid
    black prent;"
And proudlier throbb'd his heart, by far, when that same
    book he prest,
Than if a coronet and star had deck'd him—brow and
    breast.

IV.

The glass revolves again, and lo, Edina fair appears,
And men around him come and go, and Rank a proud
    front rears;

And Wealth and Fashion, gaily deck't, look on with lofty
    eye ;
While Learning, with a vague respect, bows as the Bard
    goes by.
Mark him, ye great ! The plough and clod befit him ill,
    I trow,—
The living autograph of God flash'd from his eye and brow.

V.

The drama hurries on. The bard retires to Ellisland ;
At plough and hairst-rig toiling hard—a toiler strong and
    grand !
A curb'd Elijah, peasant born, daring Song's windy height,
His homely garb clay stain'd and worn—a prophet's robe
    of light !
His giant heart his only lyre, Love's rich breath oft it stirr'd,
Till memory's passion-gusts of fire amongst its chords were
    heard ;
And never from Æolian wires was holier music wrung
Than what his heart's re-kindled fires at "Mary's" grave
    shrine flung.

VI.

The veil uplifts once more, and now, sublimest scene of all !
His lion-heart still strong, his brow erect, although the gall
And bitterness of trampl'd hopes sadden his weary soul,
As he—a stricken song god—gropes towards the final goal.
Dumfries ! no longer doth he tread the stony streets soul-
    tried—
The dark clouds settling o'er his head by genius glorified !
Ring down the curtain ! bow the head ! The last sad
    scene is o'er !
A nation mourns the mighty dead, and weep the wrongs
    he bore !

VII.

Sun, that no shadow now can cloud! Heart, that no sorrow wrings!
Man, in whose praises all are loud! Voice, that for ever sings!
A people's love, the holy bier, that holds thy worth in trust,
With glory flashing through the tear that drops above thy dust;
O rich inheritor of fame, rewarded well, at last,
Whose strong soul, like a sword of flame, smites with fierce light the past,
This sculptur'd pile, in trumpet tones, attests thy vast renown,
A nobler heirship than the thrones to princes handed down.

Stanzas for the Burns Festival

At Ayr, 1844.

DAVID MACBETH MOIR (DELTA).

STIR the baal-fire, wave the banner,
   Bid the thundering cannon sound—
Rend the skies with acclamation,
   Stun the woods and waters round—
Till the echoes of our gathering
   Turn the world's admiring gaze
To this act of duteous homage
   Scotland to her poet pays.
Fill the banks and braes with music
   Be it loud and low by turns—
This, we owe the deathless glory,
   That, the hapless fate of BURNS !

Born within the lowly cottage
   To a destiny obscure,
Doom'd through youth's exulting spring-time
   But to labour and endure—
Yet despair he elbowed from him ;
   Nature breathed with holy joy

THE BURNS FESTIVAL.

In the hues of morn and evening
   On the eyelids of the boy :
And his country's genius bound him
   Laurels for his sun-burn'd brow,
When inspired and proud she found him,
   Like Elisha, at the plough.

On, exulting in his magic,
   Swept the gifted peasant on—
Though his feet were on the green sward,
   Light from heaven round him shone ;
At his conjuration demons
   Issued from their darkness drear ;
Hovering round on silver pinions,
   Angels stoop'd, his songs to hear ;
Bow'd the Passions to his bidding,
   Terror gaunt, and Pity calm ;
Like the organ pour'd his thunder,
   Like the lute his fairy psalm.

Lo, when clover-swathes lay round him,
   Or his feet the furrow press'd,
He could mourn the sever'd daisy,
   Or the mouse's ruined nest.
Woven of gloom and glory, visions
   Haunting throng'd his twilight hour,
Birds enthrall'd him with sweet music,
   Tempests with their tones of power.
Eagle-winged, his mounting spirit
   Custom's rusty fetters spurn'd ;
Tasso-like, for Jean he melted,
   Wallace-like, for Scotland burn'd !

Scotland !—dear to him was Scotland,
  In her sons and in her daughters,
In her Highlands—Lowlands—Islands—
  Regal woods, and rushing waters ;—
In the glory of her story,
  When her tartans fired the field—
Scotland ! oft betrayed, beleagur'd—,
  Scotland ! never known to yield !
Dear to him her Doric language—
  Thrill'd his heart-strings at her name ;—
And he left her more than rubies
  In the riches of his fame.

Sons of England ! sons of Erin !
  Ye who, journeying from afar,
Throng with us the shire of Coila,
  Led by BURNS's guiding-star—
Proud we greet you ; ye will join us,
  As, on this triumphant day,
To the champions of his genius
  Grateful thanks we duly pay.
Currie, Chambers, Lockhart, Wilson,
  Carlyle,—who, his bones to save
From the wolfish fiend, Destruction,
  Couch'd like lions round his grave.

Judge not, ye whose thoughts are fingers,
  Of the hands that witch the lyre ;
Greenland has its mountain icebergs,
  Ætna has its heart of fire :
Calculation has its plummet ;
  Self-control its iron rules :

Genius has its sparkling fountains;
  Dulness has its stagnant pools.
Like a halcyon on the waters,
  BURNS'S chart disdain'd a plan:
In his soarings he was heavenly,
  In his sinkings he was man!

As the sun from out the orient
  Pours a wider, warmer light,
Till he floods both earth and ocean,
  Blazing from the zenith's height;
So the glory of our poet,
  In its deathless power serene,
Shines—as rolling Time advances—
  Warmer felt and wider seen:
First Doon's banks and braes contained it,
  Then his country form'd its span;
Now the wide world is its empire,
  And its throne, the heart of man!

Wild Flowers from Alloway and Doon.

ALEXANDER ANDERSON.

No book to-night ; but let me sit
And watch the firelight change and flit,
And let me think of other lays
Than those that shake our modern days.
Outside, the tread of passing feet
Along the unsympathetic street
Is naught to me ; I sit and hear
For other music in my ear,
That, keeping perfect time and tune,
Whispers of Alloway and Doon.

No scent of wither'd flowers has brought
A fresher atmosphere of thought,
In which I make a realm and see
A fairer world unfold to me ;
For grew they not upon that spot
Of sacred soil that loses naught
Of sanctity by all the years
That come and pass like human fears?
They grew beneath the light of June,
And blossom'd on the banks of Doon ;
The waving woods are rich with green,
And sweet the Doon flows on between ;

The winds tread light upon the grass,
That shakes with joy to feel them pass;
The sky in its expanse of blue,
Has but a single cloud or two;
The lark, in raptures clear and long,
Shakes out his little soul in song.
But far above his notes, I hear
Another song within my ear,
Rich, soft, and sweet, and deep by turns,
The quick, wild passion-throbs of BURNS.

Ah, were it not that he has flung
A sunshine by the songs he sung
On fields and woods of "Bonnie Doon"
These simple flowers had been a boon
Less dear to me; but since they grew
On sacred spots which once he knew,
They breathe, though crush'd and shorn of bloom,
To-night within this lonely room,
Such perfumes, that to me prolong
The passionate sweetness of his song.
The glory of an early death
Was his; and the immortal wreath
Was woven round brows that had not felt
The furrows that are roughly dealt
To age; nor had the heart grown cold
With haunting fears that, taking hold,
Cast shadows downward from their wing,
Until we doubt the songs we sing.
But his was lighter doom of pain
To pass in youth, and to remain
For ever fair and fresh and young
Encircled by the youth he sung.

And as to me these simple flowers
Have sent through all my dreaming hours
His songs again, which, when a boy,
Made day and night a double joy;
Nor did they sink and die away
When manhood came with sterner day,
But still, amid the jar and strife,
The rush and clang of railway life,
They rose up, and at all their words
I felt my spirit's inner chords
Thrill with their old sweet touch, as now,
Though middle manhood shades my brow.
For though I hear the tread of feet
Along the unsympathetic street,
And all the city's din to-night,
My heart warms with that old delight,
In which I sit and, dreaming, hear
Singing to all the inner ear,
Rich, clear, and soft, and sweetly turns,
The deep, wild passion-throbs of BURNS.

Robert Burns.

(On the Inauguration of the Burns Monument, Kilmarnock, 1879.)

ALEXANDER ANDERSON.

Ho, stand bare-brow'd with me to-day, no common name we sing,
And let the music in your hearts like thunder-marches ring;
We sing a name to which the heart of Scotland ever turns,
The master singer of us all—the ploughman ROBERT BURNS.

How shall we greet him as he stands a beacon in the years?
With smiles of joy and love, or bursts of laughter and sweet tears?
Greet him with all—a fitting meed for him who came and wove
Around this lowly life of ours the spells of song and love.

What toil was his! but, know ye not, that ever in their pride
The unseen Heaven-sent messengers were walking by his side;
He felt their leaping fire, and heard far whispers shake and roll,
While visions, like the march of kings, went sweeping through his soul.

"Thou shalt not sing of men," they cried, "girt up in sordid life,
Nor statesmen strutting on the stage their hour of party strife,
Nor the wild battle-field where death stalks red, and where the slain
Lie thicker than in harvest fields the sheaves of shining grain.

"Sing thou the thoughts that come to thee, to lighten up thy brow,
When, with a glory all around, thou standest by the plough;
Sing the sweet loves of youth and maid, the streams that glide along,
And let the music of the lark leap light within thy song.

"Sing thou of Scotland till she feels the rich blood fill her veins,
And rush along like sudden storms at all thy glorious strains;
A thousand years will come and pass, but loyal to thy claim,
For ever in her heart shall glow the Pharos of thy fame."

He came, and on his lips lay fire that touch'd his fervid song,
And scathed like lightning all that rose to skulk behind a wrong;
He sung, and on the lowly cot beside the happy stream
A halo fell upon the thatch, with heaven in its gleam.

And love grew sweeter at his touch, for full in him there lay
A mighty wealth of melting tones, and all their soft, sweet sway;
He shaped their rapture and delight, for unto him was given
The power to wed to burning words the sweetest gift of Heaven.

O blessings on this swarthy seer, who gave us such a boon,
And still kept in his royal breast his royal soul in tune !
Men look'd with kindlier looks on men, and in far distant lands
His very name made brighter eyes and firmer clasp of hands.

The ploughman strode behind the plough, and felt within his heart
A glory like a crown descend upon his peaceful art ;
The hardy cottar, bare of arm, who wrestled with the soil,
Rose up his rugged height, and blessed the kingly guild of toil.

And sun-brow'd maidens in the field, among the swaying corn,
Their pulses beating with the soft delight of love new born,
Felt his warm music thrill their hearts, and glow to finger tips,
As if the spirit of him who sang was throbbing on their lips.

What gift was this of his to hold his country's cherished lyre,
And strike, with master hand, the chord's of passion's purest fire !
Say, who can guess what light was shed upon his upturn'd brow,
When in the glory of his youth he walk'd behind the plough !

What visions girt with glorious things, what whispers of far fame,
That down the ladder of his dreams like radiant angels came !
What potent spells that held him bound, or swift, and keen, and strong,
Lifted to mighty heights of thought this peasant-king of song !

Hush ! think not of that time when Fame her rainbow colours spread,
And the cool rustling laurel wreath was bound about his head ;
When in the city 'mid the glare of fashion's luring light,
He moved—the moment's whim of those that wished to see the sight.

Oh, heavens ! and was this all they sought? to please a passing pride,
Nor cared to know for one short hour this grand soul by their side ;

But shook him off with dainty touch of well-gloved hand,
    and now—
Oh ! would to God that all his life had been behind the
    plough !

And dare we hint that after this a bitter canker grew,
That all his aspirations sunk, and took a paler hue ;
That dark and darker grew the gloom, till in the heed-
    less town
The struggling giant in his youth heart-wearied laid him
    down.

What were his sad earth-thoughts in that last hour—ah,
    who can tell ?
When by the pillar of his song our laurell'd Cæsar fell ?
We ask but questions of the Sphinx ; we only know that
    death
Unclasp'd his singing robes in tears, but left untouch'd
    the wreath.

Thou carper ; well we know at times he sung in wilder
    mirth,
Until the mantle of his song was trailing on the earth ;
But not for thee to lift thy voice, but leave the right to
    Heaven
To judge how far this soul has dimm'd the splendours it
    had given.

For us who look with other eyes, he stands in other light,
A great one with his hands upheld through shadows to
    the right,

Who, though his heart had shrunk beneath the doom
 that withers all,
Still wove a golden thread of song to stretch from cot to
 hall.

And now, as when the mighty gods had fanes in ancient
 days,
And up to carven roof-work swept great storms of
 throbbing praise,
So we to all, as in our heart, this day with tender hand
Uprear the marble shape of him, the Memnon of our land.

And sweeter sounds are ours than those which from that
 Memnon came,
When the red archer in the east smote it with shafts of
 flame ;
We hear those melodies that made a glory crown our
 youth,
And wove around the firmer man their spells of love and
 truth.

And still we walk within their light—a light that cannot
 die ;
It shines down from a purer sun and from a brighter sky ;
It crowns this heaven-born deputy of Song's supremest
 chords,
And leaps like altar-fire along !his deep and burning
 words.

Lo ! pause and for a moment take the seer's keen reach
 of ken,
And see the dim years struggling up with crowds of
 toiling men ;

They, too, will come, as we this hour, with passionate
    worship wrung,
And place upon those white, mute lips, the grand, great
    songs he sung.

Ho! then, stand bare of brow to-day, no common name
    we sing,
And let the music in your hearts like thunder-marches
    ring;
We sing a name to which the heart of Scotland ever
    turns,
The master singer of us all, *our* ploughman—ROBERT
    BURNS.

Burns' Vision of the Future:
A CENTENARY POEM.
(25th January, 1859.)

MILES MACPHAIL.

SCENE—*The Banks of the Nith—Midnight.*

THE moon looked from a cloudless sky
  On hill and dale, on bank and brae ;
The stars were glistening bright on high,
  And earth in peaceful slumber lay.

Oppress'd with grief, the poet mused,
  By many a doubt and care distress'd,
On time and talent oft abused ;
  And then his burning thoughts express'd.

" I've followed many a devious way,
  And heedless been by passion driven :
Alas ! the light that led astray,
  Was not the glorious light from Heaven.

" Yes ! I have erred with right in view
  Against my judgment and my will ;
The judge who holds the balance true,
  Will rightly weigh the good and ill.

BURNS' VISION.

" When Life's brief day is past and gone,
 And death's dark night has come to me,
My native land ! O ! guard thine own
 I leave my name to Time and Thee !

" I've taught man to respect himself,
 Whate'er his lot or his degree,
And not depend on worthless pelf,
 But on his own nobility.

" No more shall rank look slightingly
 Upon his low and poorer brother;
The Soul's the man !—and surely he
 Who made the one, did form the other.

" The Hypocrite I have unveiled,
 And lashed the solemn Pharisee,
But ne'er at true religion railed,
 The 'Cottar's Night' will plead for me.

" I know when Spring with sun and shower
 Unfolds the early daisy's e'e,
The thoughtful mind will bless the flower,
 And kindly, fondly think of me.

" When friends long parted meet again,
 In princely hall, or lowly cot,
Their hearts will echo back the strain,—
 'Should auld acquaintance be forgot.'

" The scenes I love shall love command,
 And warbled be in many a tune ;
Yes ! pilgrim feet from distant land
 Will tread the bonnie banks o' Doon.

"Auld Alloway's kirk, the brig o' Ayr,
    Will thousands lure the scene to see,
Where Tam o' Shanter and his mare
    Astounded were with witcherie.

"Nor be forgot, the Catrine woods
    That wanton in the summer's smile,
The crystal streams, the sparkling floods
    That kiss the braes o' Ballochmyle.

"The peasant in the furrowed field,
    Will hold his manly head full high,
His free-born thoughts he will not yield,
    Nor ever stoop to tyranny.

"I've sown the seed, and time and birth
    Will bring forth in an after age,
When Freedom's voice, o'er all the eart
    Shall be man's glorious heritage!

"On wings of fire, the world 'twill span,
    In foreign lands they'll breathe it,
'Tis not the coat that makes the man,
    But the heart that beats beneath it.

"And honest worth shall raise its heel
    'Gainst tyrants, kings, and a' that,
Till every despot learns to feel,
    'A man's a man for a' that.'"

He paused—a cloud passed o'er his brow,
    He muttered words of bitter scorn,
"Courted and slighted, praised, and now—
    Yea, surely, man was made to mourn.

His kindling eye, the tears scorched up,
    He felt ashamed a drop to see ;
" Though I have drained life's bitter cup,
    I'll live in future history.—

" When ye, with nought, but wealth can give,
    Who pass me on the other side,
Will leave no name,—or only live,
    On sculptured tombstone, false as pride ! "

He started back, a figure stood
    Whom once in vision he had seen ;
Now she assumed a loftier mood,
    A sterner and a haughtier mien.

Her brow was with the holly bound,
    A tartan plaid was o'er her thrown,
She spoke—and as he looked around
    Both hill and dale were fled and gone !

" You wished, you hoped, and breathed a prayer,
    That future times would mark your name ;
A deathless chaplet you shall wear,
    Your country will protect your fame !

" As ruthless Time on Lethe's shore
    Sweeps all into that silent sea,
He'll drop his scythe, and pass thee o'er—
    He will not touch, or injure thee.

" All hail ! my own illustrous son !
    A far-off morn I will unveil—
That glorious future is thine own,
    Behold it !—'Tis no idle tale ! "

BURNS' VISION.

He look'd and saw a Palace fair,
   It bore no crown or coronet;
No gloomy tower of stone was there,
   No martyrs' blood it wore on it;

Its crystal domes caressed the sky,
   The stars in wonder gazed to see
The fairest Fane ere raised on high
   Since earth arose from out the sea.

" Six thousand years of strife and blood—
   The Tyrant's rod—the nation's cries;
Yet,—there the People's Palace stood,
   I knew it would one day arise!

" Art shone upon its crystal walls,
   And Science all her trophies hung,
Its ancient and historic halls
   Shrined all that ere was heard or sung."

He gazed upon the rolling crowd,
   While loud hurrahs broke on his ear:
" Some victory gained," he said aloud,
   " A nation's thanks,—a British cheer!"

Fame grasped him by the hand, and said,
   " Thou long'st to know what that may be,
My son!" exclaimed the enraptur'd maid,
   " Behold—Thine own Centenary."

'Mid shouts and cheers—the curtain fled—
   He gazed with awe and joy by turns,
A laurel crown entwined—the head—
   The sculptured form was—ROBERT BURNS.

The Poet's Jubilee.

Written for Burns' Centenary, January 25, 1859, the year of Diodati's Comet.

THOMAS C. LATTO.

> Crescit occulto velut arbor ævo
> Fama *Roberti;* micat inter omnes
> SIDUS ARATRUM, velut inter ignes
>   Luna minores.

BEHOLD ! his mighty Cycle filled,
  The Scottish " Orb of Song " arise,
Return'd from the far athery wastes
  To gild our Western skies.
A hundred years ago the spark
  Stream'd upwards from a Scottish hearth,
Whose lustre now lights continents
  That wot not of his birth.

When first his march sublime began,
  Cold was the look from Scottish seers
Who flock'd to gaze upon the star
  Whose brightness all outpeers ;
But not the brightness to admire,
  Tho' wondrous radiances combine ;
Their telescopes were raised to find
  The spots that dimm'd its shine.

Deem'd they, these wise men of the East,
  The spots would flit and disappear,
Lost in the golden aureole
  Increasing year by year?
For as he circled on his course
  Gathering a glory in his train,
A host of lesser fires were doomed
  To vanish from the wain.

What, in her vast eccentric round
  Has the great Comet, COILA, seen?
What greetings, in the stellar depths,
  With compeers bright have been?
She pass'd upon the verge of space
  Anacreon's light and airy car;
And Pindar's glowing chariot wheels
  Saluted from afar.

She smiled as Horace waved his head,
  In a "fine phrenzy" onward driven,
Each showering from his golden urn
  The Poetry of heaven.
She wheeled in triumph thro' the lift
  As old Dunbar went singing by,
And the "Lord Lyon King at Arms"
  Rose rampant in the sky.

The "Gentle Shepherd's" vocal sphere
  With happy influence fill'd the air,
And Michael Bruce's twinkling light—
  No lovlier cresset there.
And beams bright as the morning star
  Flash'd from that Orb in heav'ns arcade,
When Fergusson's asylum ark
  Its proud obeisance paid.

Nor would the pulsing heart within
  But with a throb of rapture bound
As the great planet " Waverley "
  Sail'd thro' the vast profound,
And recognize with warmest beat
  The Bard of Hope as nigh he drew,
When Campbell's harness'd steeds of fire
  Swept thro' the liquid blue.

All hail! immortal ROBIN, hail!
  Thy natal day again returns,
And here the gather'd clans are met
  To crown their Poet—BURNS!
A crowded century has pass'd,
  Auld Scotia, a repentant dame,
Kneels at her ploughboy's feet and gives
  A hundred years of fame.

How dear to her his memory now—
  See to his grave how pilgrims wend;
Dear are his haunts, the fields he ploughed
  And every line he penn'd.
Rare bards have borne across the deep
  The wild rose pluck'd from Alloway's aisle,
Sprays from the birks of Ellisland
  And braes o' Ballochmyle.

O world! thy strange and rugged ways,
  Where beggars trudge the flinty road,
Refused a crust,—the beggar dies;
  The beggar is a god.
His very dust is canonized,
  What name than Homer's sounds so **grand,**
And yet the " *awmous dish* " he bore
  Foot-weary thro' the land.

Ah! twas in Greece, ungrateful State,
    That crushed her bravest in the dust,
Incensed to hear her patriot's called
    Eternally "the Just;"
That gave the cup to Socrates
    And made Miltiades to feel
The sword that waved o'er Marathon
    Was only worthless steel.

Alas! for glory and alas!
    For the illustrious Northern land,
Where once this proud Elisha stalked,
    Starving, proscribed and bann'd.
Hunger, neglect and want and woe
    He met with bursting heart alone;
He asked for work, he asked for bread,
    And Scotland gave—a stone.

He, the immortal peasant-boy
    Who shed the light of song and smiles,
Cheering alike Siberian wastes
    And fair Pacific isles:
Who, with a pathos, all his own
    Could lull the fever'd soul to sleep,
And win the banks o' Bonnie Doon
    In sympathy to weep:

Who look'd his fellows in the face
    Whate'er their wealth or name or rank,
Pouring derision on the curs
    That snarled at his flank:
Who, with a mood for every mind,
    Nerved kilted warrior for the strife,
Soothing the woes, hymning the joys
    Of simple country life:

THE POET'S JUBILEE.

Who flung a halo round the brows
    Of milkmaids lingering by the stream,
And answered with his bold black eye
    Love's own peculiar gleam :
Who set in an eternal dew
    The gowan of the grassy holm,
And reared to Freedom in the earth
    Her fairest, stateliest dome ;

Whose sympathetic heart of hearts
    Shelter'd a mouse with tenderest care,
Breathing his malison on him
    Who maimed the timorous hare :
Who snatched the mirror from the shelf,
    Set it before a peasant's eyes,
And bade him see a manhood there
    To which his soul might rise :

Who clasp'd within the arms of love
    The maid dear to his bosom's core,
Lapping her in the folds of song,
    A joy for evermore :
Who first laid down the electric cords
    That million hearts in one combine
When flash along the magic words
    He uttered—AULD LANGSYNE !

Years have roll'd on since in a dream
    I saw the poet, pale with care ;
For I have also lain and dreamed
    Beside the winding Ayr.
And a deep voice that thrill'd my soul
    Came through the stately elms around
As if some Druid struck in wrath
    A harp of solemn sound :

" Forbear, fond youth, with holly wreath
    To twine thy young, exultant brow
And learn, ere yet too late, the doom
    Of dreamers such as thou :
Days of distress and nights of pain,
    Moments of bliss and months of gloom,
To gild thy memory a feast,
    And to thy shade a tomb.

" But, if the lava in thy veins
    Must issue in the stream of song,
Then whet thy sword and dart thy lance
    At mean pretence and wrong.
Stand up against Oppression's host,
    With dauntless eye confront their frown,
And if a tyrant cross thy path
    Then hew that tyrant down.

" If thou should'st find a noble heart
    Too proud to sue, too strong to bend,
Then let that heart be knit to thine,—
    Then make that man thy friend,
Heed not the clamour of the crowd,
    True to the promptings of thy soul,
Let the inexorable right
    For ever be thy goal.

" But know if but one hairbreadth's track
    From Wisdom's path thy footstep stray,
The hell-hounds panting for thy fall
    Shall dog thee on thy way.
Proud in your manhood as you stand
    Thou must be cabined and confined,
For Boards and Lords have still the power
    To cheer the immortal mind.

"I, too, was proud; I, too, was tamed;
    The Autocrats of Scottish drink
Could warn the Gauger, with a sneer,
    To act and not to think.
I loved the Thistle well, God knows;
    For Scotia's sake year after year,
I turned my weeder-cleps aside
    And spared the symbol dear.

"I sung the Bruce and 'Wallace wight,'
    Rekindled smouldering martial fires,
Glowing with pride to trace my line
    From such unconquer'd sires.
But yet a stepdame, grim and cold,
    She frown'd rebuke, she chill'd my glee;
For what, forsooth, was 'Wallace wight'
    Or all his chiefs to me!

"Tears flowed as to the mercy-seat
    The Cottar's artless prayer was borne;
But nations joined the echoing wail
    That 'Man was made to mourn.'
Still, it was something to have lived
    And felt, that 'neath the Godhead's smile,
Affliction's furnace tries like gold
    The noble sons of toil!"

The vision vanish'd, and the voice
    Of Wisdom ceased its words of love,
And I awoke to know and feel
    That dreams are from above.
Hark! now the Jubilee trumpet sounds
    Through the wide world its loud acclaim,
Drowns the remembrance of his wrong
    And of a country's shame.

Yes! "brither Scots," arise, rejoice!
    Drink, eat, be merry as ye may;
Sadness becomes no Scottish face
    On BURNS's natal day.
And, Scotia! as the wine-cup shines,
    And as thy stubborn brows unbend,
Bethink thee of the chance thou miss'd
    And "tak a thocht an' mend."

Perchance this hour within thy vales
    Some suffering child of song may lie,
Whose bleak existence may be life
    So far—he does not die.
If, timely uttered, one kind word
    Might cheer and raise him from the dust;
Scotland! thou ow'st the Bard a debt,
    Be generous and just!

Robert Burns.

W. Stewart Ross.

All hail, O Nithsdale's furrowed field, a Marathon art thou;
The fire of God in his great heart, of genius on his brow,
Thy patriot bard strode o'er thy sward, his triumph car the plough!
The laverock in the early dawn, the merle at evening grey,
Sang pæans as the ploughman trod his more than laurelled way,
And the red ridge of Scottish soil behind him grandly lay,
Prinked with the daisy's "crimson tip," the "rough burr-thistle's" head,
And rough print of the ploughman's shoe—shoe of the deathless dead.

'Tis o'er, the rig is dark with night, "the "lingerin' stars" on high,
And song-land's gained another wreath of flowers that never die.

In Nithsdale, as a dreamy boy, in wild ecstatic turns
I've grasped the plough to follow wrapt thy shade, O Robert Burns!

As "spretty nowes have rairt an' raikit," I've seen thee standing nigh,
'Mid visions of the throne of song too grand for mortal eye:
The hills around burned into verse an anthem vast and dim,
The "fragrant birk" an idyll grew, the "stubble field" a hymn!

O sword, rust o'er thy mighty dead, pent in their funeral urns,
Plough, by Elisha sanctified * and glorified by Burns,
Thine is no roll of tears and groans, the dying and the dead,
Thou writest on the wintry field the prophecy of bread :—
I'll drive my share o'er vanquished want, my coulter's edge uprears
The banners of the yellow corn, the rye's unnumbered spears.
God speed thy "horns"—no altar horns so sacred are to me,
The Prophet † and the Muse ‡ of Fire their mantle bore to thee!

Yet, would a tyrant weld our chains? Then victory or the grave—
The trumpet blast of "Scots wha hae," will make the coward brave!
Then onward, valour "red-wat-shod,"—glory to him who dies!
Be his eternal infamy, the "traitor knave" who flies!

---

* "The Poetic Genius of my country found me, as the prophetic bard Elijah did Elisha, at the plough, and threw her inspiring mantle over me."—BURNS, *in dedication to Second Volume of his Poems.*

† *Elijah*, frequently and aptly designated the Prophet of Fire, from fire being so constantly the *sign* in his divination.

‡ *Coila*, the Muse of Burns, peculiarly the poet of the fire of passion and poetic ardour.

Dumfries, thy cold hands hold his urn, thou guard'st his iron
    sleep,
O shrine that draws the universe to worship and to weep !
What tribute grand of brass or stone can thy poor hands
    bestow ?
What bronze or marble worthy him who lies so cold and
    low?
Of the brave man whose fight is fought, whose weapon's
    sheathed, whose banner's furled,
Tho' still his fire and force of soul throb in the veins of half
    the world :
Australia loves him, India too, as tho' he had but died
    yestreen ;
Columbia knows the "Banks o' Doon," and Afric sings of
    "Bonnie Jean ! "

Hast seen athwart the midnight stars a cloud its shadows
    fling?
Hast seen the stain from cage's bars upon the eagle's wing?
Impeach I will not ; but, Dumfries, I cannot do him wrong.
Thy street-mire stained the singing robes of the great king
    of song :
Look sorrowing back on the grey hairs too early o'er his
    brow,
And, grateful, what he lacked in bread give him in garlands
    now :
Humble am I who ask you this penitence be done,
But, O city of St. Michael,* remember I'm thy son ;

I love thee tho' I'm far away, tho' you've forgotten me,
My dreams of home and fatherland are centred all in thee ;

---

* Dumfries, of which St. Michael was patron saint.

I ask for nothing for myself, I tread thy streets no more,
Honour thyself by honouring him I and the world adore;
Joy in thy solemn heritage, breaking oblivion's wave,
O grandest city of the world, for you have BURNS's grave!

To Burns.

John M'Intosh.

The Scotsman is prood o' his thistle an' brier,
   O' his heather-clad hills an' his iron-bound shore,
O' his wide-spreading forests, the hame o' the deer ;
   Ay, he lo'es them a' weel to his innermost core.
But ower a' there's ae day i' the dawn o' the year,
   Whaure'er he may be, like a pilgrim he turns
To gaze thro' the mist o' affection's fond tear
   On the laurel-crowned name o' oor ain Rabbie Burns.

       Chorus—
   For he was the laddie wha sang at the plough
      As lichtsome a lilt as the lark's cheery strain,
   An' aye on his birthday oor pledge we renew
      Oor auld mither tongue in her pith to maintain.

We hae Makkars 'mang kings, amang lawyers, an' priests ;
   There was Jamie, an' Barbour, and douce Quintin Schaw,
A' men wi' the true Doric fire i' their breists ;
   Oh, we ken them richt weel, an' are prood o' them a' ;
But the years slip awa', wi' their fasts an' their feasts,
   An' tho' aye in due order each Birth-day returns
We've nae mind o' the tane till 'tis far by the neist's,
   But we couldna dae that wi' oor ain Rabbie Burns

TO BURNS.

Chorus—
  For he was the laddie wha sang at the plough
    A lilt that's been learnin' to lassie an' swain,
  An' dull maun the cuif be wha winna aloo
    That love-liltin' RAB was a freen o' his ain.

The auld-warld bodies they warna a' fools,
    Although aboot them I've nae business to speak,
As for soondin' their names, sirs, I haena the rules,
    An' their warks are to maist o' us sae meikle Greek;
But we a' ken this much, that when laid in the mools,
    Or burnt to wee bittocks and stappit in urns,
Their sangs were but kent to a wheen o' the schools,
    But it never was sae wi' oor ain RABBIE BURNS.

Chorus—
  Na, na, for the laddie wha sang at the plough,
    It wasna to schoolmen he liltit his strain,
  But to lichten the cloud on each pale toiler's broo,
    An' to heeze him to heichts as sublime as his ain.

Then, Scotsmen, where'er your assembly may be,
    Aye mind oor auld tongue can be never forgot,
Till the 'oor whirls roon' when Eternity's e'e
    Gies the wink to auld Time to be aff like a shot.
An' far through the vista o' the ages I see
    That where manhood exults or misery mourns,
They'll hae balm to their waes, they'll hae gust to their glee,
    In the life-breathin' sangs o' oor ain RABBIE BURNS.

Chorus—
  Then here's to the laddie wha sang at the plough,
    Come join ane an' a' in the hearty refrain;
  An' foul fa' the dullard wha winna aloo
    That we never shall look on his marrow again.

The Soul of Burns in Song.

HON. WILLIAM C. STUROC.

OH Muse! come once again,
  Tho' far from Scotia's shore,
Inspire this senile brain,
  With Fancy's fire once more;
And ere life's work is done,
  This latest tribute be
A wreath for Scotia's son—
  Great BURNS, I'd sing of thee!

Thy birth, thy life, thy death,
  A thousand tongues have told;
Nor mine can add a breath
  To fame like burnish'd gold,
That flashes o'er our race,
  With lightning's speed and power
And casts the Poet's grace
  On each succeeding hour;

On each succeeding year—
  O'er Cycles as they run,
Till man's mundane career
  Shall pass like setting sun;

While brighter grows thy fame,
    And Scotland's heart beats proud,
To own her BURNS' name,
    In anthems long and loud.

Nor Scotland's sons alone,
    Proclaim his matchless words,
The world's great heart doth own,
    In deep responsive chords,
That " Power," and " War," and " Wrong,"
    No more shall flout in sight ;
While BURNS' soul, in Song,
    Will mingle with the *Right!*

A Ploughman's Dream.

(Written by request for, and read to The * Brethren of The Braes, at their meeting in celebration of the 125th anniversary of the birth of Robert Burns.)

ROBERT HOGG.

"The Poetic Genius of my country found me as the prophetic bard Elijah did Elisha—at the *plough;* etc."—See Burns' "Dedication," in the "Edinburgh Edition" of his poems.

SEE yonder horny-handed son of toil,
   Of lowly station and of humble name,
Stand by his plough among the furrow'd soil,
   His eyes fixed on the far-off steeps of fame,
His yearning gaze revealeth—to aspire
To top those heights—his heart's one proud desire.

He looks around and marks a skylark rise
   From neighbouring field, where lowly lay its nest;
With quivering wing it seeks the cloudless skies,
   Scattering the dewdrops from its "speckled breast,"
Pouring meanwhile its sweetest God-given lay;
Rejoicing zephyrs bear it far away.

He lists enraptured, as it were the voice
   Of seraph hymning in the courts above.

---

* A Rambling Club.

A PLOUGHMAN'S DREAM.

" 'Tis song," he cries, "makes heavy hearts rejoice,
   'Tis song that soothes the spirit all to love.
Song scathes the Wrong and lights the path of Right;
Song nerves the patriot-arm for the fight!

" Soar soul of song! most gifted bird of birds!
    Fain would I learn all thy guileless art;
Fain would I weave thy song in living words
    That burn their way into the cold world's heart;
For now a wish doth 'strongly heave my breast,'
For Scotland's sake to 'sing a song at least.'

" Oh, cruel Fate! Why was I lowly born,
    With mind aspiring after higher things
Which keep aloof from me, and laugh to scorn
    The fitful efforts of my feeble wings?
But I will not be daunted, for I feel
A conscious power which care cannot congeal.

" A stirring in my soul proclaims the hour
    Is nigh, when men shall point at me and say—
'There goes a bard whose winged words of power
    Do breathe, and burn, and fill the mind alway
With lofty thoughts that lead to high empire;
And glorious songs flow from his living lyre!'

" Then, sing on, skylark, sing though lost to sight,
    All Nature lists heart-charmed thy melody;
Oh, that my soul could follow in thy flight!
    Oh, that my heart could pour thy wonderous lay!
Then might I win the enraptured crowd's acclaim!
Then might I tread the giddy steeps of fame!"

A PLOUGHMAN'S DREAM. 215

He wipes the perspiration from his brow,
   And sighing heavily, sits down to rest;
In musing mood he leans against his plough,
   Soon sleep-subdued, his chin falls on his breast;
And as he sleeps he dreams a sad sweet dream,
Till rudely startled by the plover's scream.

And thus he dreams :—

               He stands beside his plough
   At morn and sees a vision wonderous fair ;
A maid with lustrous eyes and noble brow,
   Long gleaming locks and snowy bosom bare ;
A sun-bright mantle wraps her lovely form,
Which could not fail the heart of man to charm.

"Ploughman," she kindly spake, "I know with toil
   Thy hands are hardened, but thy soul is soft ;
And care shall plough it, as thou dost the soil ;
   And though at Cupid's acts thou scoffeth oft,
And toy with him in many a sprightly lay,
Yet Love shall lightly bear thy heart away.

The wish that doth thy honest bosom fill,—
   Man's heart to charm with song,—thou seest I know ;
But first, with Love thine every pulse must thrill,
   And wrapt thy very being in its glow.
But, when young Cupid with thy heart hath sped,
I'll grant thee *Angel Israfel's instead."

---

   * And the Angel Israfel, whose heart-strings are a lute, and who has the sweetest voice of all God's creatures.—KORAN.

A PLOUGHMAN'S DREAM.

In panoramic view before his eye
    She then bids pass—a Highland maiden sweet,
Her stay is transient as an infant's sigh,
    Yet ere she flies his heart pants at her feet ;
Anon another fair one claims his smile,
And cheers him with her winning words the while.

The plough he follows now with measured stride,
    And many kindly words his team to cheer ;
And here the " modest " daisy is espied,
    And o'er its luckless lot he sheds a tear ;
And now he turns the " weeder-clips aside "
To spare " the rough burr-thistle," Scotland's pride.

And as he " turns a fur " with pitying smile,
    He marks the " tim'rous beastie's " dread affright ;
Cæsar and Luath grow " unco pack " the while,
    And now the " tenants of the lake " take flight ;
Anon a wounded hare limps o'er the plain—
A sight that fills his tender heart with pain.

The week's toil o'er, now seated by the fire,
    He joins the cottars in their hymn of praise ;
With reverence due, he list the grey-haired sire
    " The younkers " warn to walk in wisdom's ways ;
He sees him kneel, and pray the God of Love
" That thus they all shall meet in " heaven above.

He laughs to view thy mirth provoking face,
    " Auld Hallowe'en," the lovers' dear delight ;
Scotland's " great chieftain o' the Puddin' race,"
    " Warm, reekin', rich," now ravishes his sight ;
While Bacchus' sons declare there can't be found
" Three blyther he'rts " the known world around.

A PLOUGHMAN'S DREAM

He hears the "Souter" tell his stories "queer,"
    While "Tam o' Shanter" tipples at his side;
The landlord's "ready" laugh falls on his ear,
    He follows Tam on his immortal ride,
Sees in Kirk Alloway at dead of night,
A "hellish legion" dancing with delight.

He hears an agéd couple "hand-in-hand"
    Talk pensively of joys long passed away;
Now dawns upon his sight a jovial band
    Clad all in rags, yet laughing noisily;
He starts to see the grim gaunt form of Death;
Now Satan's self appears—he holds his breath!

He kindly greets a "brither bard an' freen,"
    And after him "a fine fat fodgel wight;"
Sees "Holy Willie" shuffle on the scene,
    With many another canting hypocrite;
Then, lo, in bands the "unco guid" repair
With faces long to "Mauchline Holy Fair."

Well pleased he wanders by his "windin'" Ayr,
    While dim and distant her "twa brigs" appear;
Then by Doon's "banks an' braes" "sae fresh an' fair;"
    Now "Scotia's darling seat" Edina dear
He hails with joy.—Here many a night and day
Go fleeting past in mirth and gaiety.

He stands uncovered by the lowly tomb
    Of brother bard, whose song to him is dear;
An agéd minstrel wails man's hapless doom;
    A syren's voice whispering into his ear
Sylvander meets the "mistress of his soul!"—
Mad passion rushes on beyond control.

A PLOUGHMAN'S DREAM.

Then in the gentle hush of * April e'en,
   At heart resolved to right the wrong, he strays,
His arm about the waist of "bonnie Jean,"
   Who eagerly lists every word he says ;
She joys to hear him old love vows renew
That unto her his heart will aye be true.

Crushed 'neath oppression's heavy hand, he sees
   The poor man fainting, failing, fight 'gainst fate ;
He hears the afflicted's moan on every breeze,
   He sees the rich man honoured ear' and late ;
And deep within his soul, he yearns that he
Might wake the Anthem of Democracy.

He sees a land of Scottish heroes kneel
   Around their leader vowing ne'er to yield ;
He sees the flower of England's army reel
   And flee in terror Stirling's fatal field !
He sees the Bruce arise, and snap in twain
The tyrant Edward's, galling, servile chain !

Beside his feet, he marks a streamlet flow,
   How dark its waters to his eyes appear !
The maiden stoops and kisses it, when, lo !
   It in the sunshine sparkles bright and clear.
Again she speaks : "Oh, Burns, thou shalt ere long
Thus purify the stream of Scottish song.

---

* Burns returned from Edinburgh about the end of March, 1788. The first allusion to his marriage which appears in his correspondence we find in a letter to his old associate James Smith, dated—"Mauchline, April 28, 1788."

"But, as the seed that sinks into the earth
　　Needs shine and shade ere its fruition sees,
So shalt thou have thine hours of thoughtless mirth,
　　And drain thy cup of sorrow to the lees;
But, as the sun outshines the stars of night,
All other bards shall pale before thy light!

"Thou'rt lowly born? Not lowlier than he
　　Whose matin lay hath oft bewitched thine ear!
Thy fare is plain? His plainer could not be,
　　Yet angels cease to sing his song to hear!
Thy raiment is but rough-spun 'hodden grey?'
And is the lark a bird of plumage gay?

"Though sweet the roses of the summer blow,
　　When autumn comes, they're scentless all and sear;
While 'modest, crimson-tippéd' daisies grow,
　　'Mid foul and fair, throughout the circling year;
And so thy song shall blossom for all time,
From age to-age, in every land and clime.

"Scotland *may* perish! Her cities crumbling fall,
　　The surging waters o'er her mountains roll;
Kings, statesmen, martyrs, heroes—one and all—
　　Together find Oblivion's gulf their goal;
E'en Scotland's honoured name, some distant age,
Time's ruthless hand may blot from history's page.

"But long as human hearts can find delight
　　In Love and Liberty, the wide world o'er;
While honest tongues shall plead the cause of Right,
　　Till man shall crush his fellowman no more;
Enshrined within those hearts shall live thy name;
Those tongues with joy sing thy undying fame.

" And low and high, the cynic and the sage,
    Shall o'er thy musings drink of joy divine ;
All men enraptured scan thy wonderous page,
    Thy lowly grave become the world's shrine ;
Earth's bards united, hail thee minstrel king,
While to thy praise their throbbing lyres ring ! "

. . . . , .

Wildly and shrill the whirling plovers scream,
    He starts up to his feet, and rubs his eyes :—
"And did I sleep ?" he says : "and did I dream ?
    Ah ! was it *but a dream ?* " he sadly sighs.
But, see, his step is firmer than of yore !
A brightness in his eye unknown before !

Kossuth at the Grave of Burns.

W. STEWART ROSS.

YOUR patriot sword, your patriot lyre,
   Thrilling the world by turns,
Beats high the soul's immortal fire,
   KOSSUTH and ROBERT BURNS.
Blest be for aye the glorious bond,
   In minstrelsy or war,
That links the patriot mind to mind
And hails, as of one kith and kind,
   The Scot and Magyar.
Blest be for aye the patriot tie
   That joins the deathless brave,
And arches with one common sky
   The Cluden and the Drave,
That living man that standeth there
   And the dead man in the grave,
The gowans on the banks of Ayr,
   The vines by Danube's wave.
And far-off Hungary's olives blend
   With Scotland's haws and slaes,
And the wild Alps their greeting send
   In love to Logan Braes.
Ye "Scots wha hae wi' Wallace bled,"
   Green be your wreaths of fame ;
The torches of the deathless dead
   Still set the world aflame ;

Down goes the tyrant's shattered throne
   And breaks his dungeon chain,
The trampled Right bursts into light—
   The land is free again!

The flag that waved in olden days
   O'er Scotland's spearmen true
Lights with its ancient glory blaze
   Carpathian mountains blue;
And the high heart and the strong arm
   The Pass of Brander* saw
Meets in the wheel of reddened steel
   The legions of Haynau.†

The Brandanes‡ were the Honveds'§ sires;
   The Hapsburg tyrant spurn,
In Austria's blood baptise Zegrad,‖
   The child of Bannockburn;

---

* The Pass of Brander, the famous dark gorge which narrows into the Pass of Awe, the scene of the desperate engagements between Wallace and the catarins of Macfadzean and Bruce and the Macdougalls of Lorn.

† General Haynau, one of the most able and active of the Austrian leaders, inflicted a crushing and final defeat on the Hungarians in 1849.

‡ *Brandanes*, the name given to the 10,000 Scottish archers, exceptionally tall and handsome men, who, in 1298, followed Sir John Stewart to the battle of Falkirk. In that fatal engagement, after which Wallace could never again bring an army into the field, Sir John Stewart was slain and the Brandanes annihilated.

§ *Honveds*—Defenders of the Home, of whom there were ten battalions organised by the Hungarian insurgents.

‖ Zegrad, where, at the head of 40,000 troops, Jellachich, ban of Crotia, crossed the Drave in 1848. He was met by the Hungarians and defeated.

And strike the chord that fires the sword
   And the valour of the brave,
Peal forth the song that girds the strong—
   The song that shakes the grave.
And Scotland's "rough burr thistle"
   Gives verve to Hungary's vine,
And the squares* of awful Carron
   To Hungary's marshalled line,
And the pibroch and the slogan,
   'Neath Albyn's stormy sky,
Peals through red Swechat's † riven lines
   The watchword, Do or die.

Hurrah for Coila's patriot bard
   And the anthems of the free!
Hurrah for Monok's ‡ hero-son
   And the knight of Elderslie!
Hurrah for history's mighty Past
   And the unborn To-be!

And Genius links Dumfries and Perth,
   The helmed and sworded strong,
Through him, in his undreaming rest,
   The Charlemagne of song,

---

* At Falkirk, on the Carron, the Scottish infantry was disposed into *schiltrons*, or hollow squares, with the spears pointing obliquely outward against the charging cavalry, adumbrating Wellington's squares at Waterloo.

† Swechat, near Vienna, where the Hungarian patriots were defeated by the Austrians under Jellachich in 1848.

‡ Monok, or Monck, where Kossuth was born on April 27th, 1806.

KOSSUTH AT THE GRAVE OF BURNS.

Who, 'neath the marble, kingly lies
  By Nith's pellucid river,
While rings the fire-lilt of his songs
  Far down the dim Forever.

Bend lowly o'er that tomb, Kossuth,
  The priceless, peerless brave;
Our pens in rust, our voices dumb,
Shall the long future's pilgrims come
  O'er mountain, plain, and wave,
To wake the harp's Tyrtæan chords,
To ban oppression's tyrant hordes,
To fire their souls, to whet their swords,
  At BURNS's hallowed grave.

On Visiting the Tomb of Robert Burns.

WILLIAM HOGG.

DEPARTED spirit, o'er thy dust
   With reverential awe I bow;
Earth never held a greater trust
   'Mong monarch minds of man below.

And while I linger o'er thy still,
   But highly honoured bed of rest,
Emotions deep, unceasing thrill
   The life strings of my brain and breast.

Thy natal morn was ushered in
   Amidst the howl of winter drear;
Bleak Winter seemed to thee akin
   In all thy weary wanderings here.

But though the wild and withering wind
   Of adverse fate against thee blew,
The breathings of thy powerful mind
   The blights of fate could not subdue.

THE TOMB OF BURNS.

Thy song-tuned soul, with giant power,
    Shone through the dark, deep clouds of gloom
That o'er thy head were known to lower,
    And brought thee to an early tomb.

Thy quick creating mind, could make
    The awful seem more awful still,
Could phantoms wild to life awake,
    And servants make them at thy will.

Thy soul was stirred to strike the lyre
    By Nature's hand, but not by art;
Thou gav'st to song a blaze of fire,
    Undying in thy country's heart.

And while the power of stately kings
    Away from mind and memory flies,
Thy fame, on admiration's wings,
    Though great in height, will higher rise.

While worth shall hold on earth a place,
    And earth upon its axis turns,
The hand of time will not efface
    Thy golden gems, immortal BURNS!

A Burns Anniversary Song.

Sung at the Annual Festival of the London Robert Burns Club, 1885.

COLIN RAE BROWN.

IN classic Coila long ago,
   Upon a brawling, blustering morn,
When Winter's wildest blast did blow,
   The King of Scottish Song was born.
And still his fame o'erflows the earth,
   From pole to pole, from zone to zone;
While his once humble place of birth
   A universal shrine hath grown!

      For BURNS is our own King of Song,
         Our crowned King of Melodie,
      No Kaiser bold with armies strong,
         Can rule the heart so well as he!

His Empire is the heart of man,
   Yes! and the heart of woman too;
The lassies' hearts he could trepan,
   And so he taught the lads to woo.

ANNIVERSARY SONG.

Why prate of Homer and such Bards
   As deified grim Butcherie ?
True fame the Bard of Doon rewards,
   Who sang of Love and Amitie !

By song he crush'd the pride of birth,
   Taught man to love his brother man,
And showed the power of sense and worth
   To glorify life's little span :
He blew a blast both clear and loud,
   He shook oppression to the core,
While honest hearts wild pulsing, vowed,
   To suffer crushing wrong no more !

The world's great heart our Bard hath won,
   It yields him kingly homage now ;
In every land beneath the sun
   Immortal holly binds his brow.
Then let us bliss his natal day,
   That dawned in Coila long ago ;
For though the man be turned to clay,
   The Poet never death shall know !

   For BURNS is our own King of Song,
      Our crowned King of Melodie,
   No Kaiser bold, with armies strong,

Lines Written in Burns' Cottage.

Rev. R. S. Bowie.

O BURNS! the matchless, deathless, and divine,
Here, in the cottage, to thy mem'ry dear,
We sit and ponder o'er that life of thine,
Which oft hath made us shed the silent tear.
O bard of Scotia! nay, of all the earth—
Here pilgrims from all lands together meet,
To do obesience at the shrine of worth;
Here strangers rest and hold communion sweet
With those ne'er known before, because of thee!
O how thy songs can melt auld Scotland's face,
And make in her sons their brothers see.
Ay, e'en the flowers that bloom on Doon's sweet braes,
Are loved and treasured for the Poet's sake,
And in our hearts their best emotions wake!

At the Grave of Robert Burns:
A Remonstrance with some of his Biographers.

CHARLES MACKAY, LL.D.

I.

Let him rest; let him rest;
The green earth on his breast;
And leave, oh! leave his fame unsullied by your breath.
Each day that passes by,
What meaner mortals die,
What thousand rain drops fall into the sea of death.

No vendor of a tale
(His merchandise for sale)
Pries into evidence to show how mean were they.
No libel touches them;
No curious fools condemn;
Their human frailties sleep for God, not man, to weigh.

And shall the Bard alone
Have all his follies known—
Dug from the misty past to spice a needless book—
That envy may exclaim
At mention of his name—
The greatest are but small, however great they look!

II.

Let them rest their sorrows o'er,
All the mighty bards of yore ;
Or if, ye grubbers up of scandals dead and gone,
Ye find amid the slime
Some sin of ancient time,
Some fault, or seeming fault, that Shakespeare might have done,

Some spot on Milton's truth,
Or Byron's glowing youth,
Some error not too small for microscopic gaze—
Shroud it in deepest gloom,
As on your father's tomb
You'd hush the evil tongues that spoke in his dispraise ;

Shroud it in deepest night,
Or, if compelled to write,
Tell us the inspiring tale of perils overcome,
Of struggles for the good,
Of courage unsubdued,
But let their frailties rest, or on their faults be dumb.

To the Memory of Robert Burns.

On unveiling his Statue at Dundee.

WILLIAM REID.

YE who revere the Poet's hallowed shrine,
   Look on this statue with exulting eyes,
And see the glory of his fame entwine
   Around the memories you love to prize ;
And as ye gaze upon his manly form—
   Here moulded by the classic hand of Art,
Let fervent homage every bosom warm
   To greet the cherished Poet of the heart.

The lineaments unveiled before you now
   Revive enchantment's retrospective dream,
When Heaven enkindled on that noble brow
   A spell that wakes a never-dying theme ;
Swayed by its influence a nation's pride
   Rejoicing owns the genius that recalls
The sorcery of him—whose fame, earth wide,
   Crowns song with glory in life's festive halls.

Oh ! Scotia ! nurse of his poetic fire !
   We hail thee in the rapture of his strains,

And trace thy spirit in his deathless lyre,
   Which thrills the echoes of thy loved domains,
And wreathes thy beauty in immortal lays—
   Where truth and feeling brighten as they flow ;
While the impassioned fervour of their praise
   Enshrines thee with a patriotic glow.

Song's noblest tribute thou hast won from him,
   Who burned to serve thee, and exalt thy name,
And, with a lustre that shall ne'er grow dim
   Made thy fair realm the temple of his fame.
While thine the glory and immortal boast
   Of ushering his regal worth to light,
Not called to lead a sanguinary host,
   But wear a crown of intellectual might.

Amidst the constellated peers of mind
   Can genius claim a brighter soul of song ?
Or in the myriad ranks of human kind
   What heart e'er glowed with sympathies as strong ?
Not even Shakespeare's world-encircling name
   Can pour a glory where his may not shine ;
Nor rouse enthusiasm's vital flame
   To pale the lustre round thy Poet's shrine.

Attest it, ye who traverse distant lands—
   From Nova Zembla to Australian shores,
Or him who haply tracks Arabian sands,
   Or Afric's burning continent explores ;
'Midst Scythian wilds, o'er plains of Hindostan,
   Throughout all countries—under every clime,
His genius is a glorious talisman,
   Enkindling friendship, cherished for all time.

His Doric strains entrance the dreaming ear,
　And breathe Æolian rapture in the soul—
Soft warbled in an Orient atmosphere,
　Or where the sun-god gilds the western goal;
Wherever man pursues the aims of life,
　Or social ties inspire the human breast—
There, like a sunburst through the tempest strife,
　His spirit shines—an aureole of the blest.

Although no grandeur heralded his birth,
　Yet in the shade of poverty's low vale,
Fond Nature guarded his advent on earth,
　And Poesy, in secret, bade him hail;
Whilst Scotland, gazing on her wond'rous son,
　Watched his young raptures with prophetic eyes—
Beheld him reap his triumphs—all self-won,
　Then proudly claimed him with heartfelt surprise

But ah! too late to snatch him from the doom,
　Which death in secret veiled within his breast;
Worn out with care and penury's sad gloom,
　That darkly agonized his soul's unrest;
Thus overcome, he sank into the grave
　Ere life's elixir reached his fervid lips,
And to his weeping country, dying, gave
　A name,—earth's proudest born will not eclipse.

He came amongst us clad in russet guise,
　Which genius changed to an imperial robe,
And with its sceptre bade the Poët rise
　To sway the soul and all its secrets probe;

He touched the heart, as with a wizard's spell,
  And burning passions rose at his command;
While Patriotism, with heroic swell,
  Leapt up death-armed to guard his native land.

And Liberty, whose lightning-gleaming eyes
  Dart all their vivid flashes through the soul,
Bade deathless scorn in every breast arise
  To wither Tyranny's abhorred control—
As visioned Glory stalked with regal Bruce,
  When "Scots wha hae" electrified his host,
Enkindling ardour during death's short truce,
  Till Valour stood immortal at its post!

Then waking up to mark life's social wrongs,
  He rose indignant to rebuke the proud,
And claimed the right which to mankind belongs—
  Not for himself alone—but all allowed.
Corruption shrank beneath his kindling glance,
  And fell Oppression felt his scathing power;
While Villany recoiled before his lance,
  And haughty Arrogance was seen to cower.

And arch Hypocrisy's detested mask
  Fell 'neath his satire, keen as pointed steel;
While Ridicule illumed his mental task
  In lashing Bigotry's malignant zeal:
'Thus in life's battle he withstood the brunt
  When fighting for Humanity the while;
And Independence reared its manly front
  Where he commingled with the sons of toil.

And lighting up his sympathetic Muse,
   He warmed the kindred hearts of young and old,
In vivid strains which, day by day, diffuse
   A social influence that is untold.
And Pity's tear and tender Pathos swayed
   The tremulous vibration of his lyre,
While Wit and Humour saliently played
   In corruscations of poetic fire.

Still softer feelings graced his deathless song,
   And woke emotions in the kindling breast
Where Love's divine sensations swept along,
   And thrilled its chords with raptures of the blest:
And Beauty still grew beautiful the more
   He showed its charms to admiration's eye,
While lovers lingered o'er it to adore,
   Or heave the soft involuntary sigh.

Thus Nature, listening to his matchless lyre,
   Smiled with delight upon her favoured Bard,
And lit his numbers with immortal fire,
   To be thereafter his supreme reward.
And Scotia's Muse, with emblematic charms,
   Entranced his vision in his humble shed—
Showed him her Sages, and her chiefs in arms,
   Amidst the glories of her deathless dead;

Her mountains, vales, her seas and shining streams;
   Her lochs, her glens, her groves, and purpled heath,
And hues, and forms, which light the Poet's dreams,
   Then crowned her minstrel's brow with her own
      wreath:

MEMORY OF BURNS.

The world beheld—and drew him to its glare—
Full in its sun to test him by its blaze ;
'Midst giants he became a Titan there—
Nor dwindled in an atmosphere of praise.

He viewed the varied lot assigned to man,
 And scanned the mysteries of human life :
Then with their lights and shades conceived his plan
 To dignify mankind and banish strife.
He felt himself the moral of his song,—
 Life's passions, errors, and regrets and shame ;
So weighed the issues between Right and Wrong,
 And scorned the cost that might redeem a name.

Thus nobly he pursued his manly task,
 Though time to ripen was denied by Death ;
What had he been if spared ? 'twere vain to ask,
 His triumphs gained now live on every breath.
Farewell, Great Bard !—and yet why say farewell,
 When to his semblance here, the vision turns ?
While memory bids the glowing bosom swell
 As pride and pleasure point to ROBERT BURNS.

Robert Burns.
(*Canada, January 25th, 1888.*)

JOHN MACFARLANE.

TO-NIGHT amid Canadian snows,
   In lordly hall and cottage home,
Where'er the blood of Scotsmen flows,
   Where'er the feet of Scotsmen roam;
ONE name upon the lips grows sweet,
   More rich than wine from purple urns,
With thrill electric flashing fleet,
   The name of ROBERT BURNS.

Young hearts through all the golden years
   Proclaim the magic of his wand,
And aged eyes are wet with tears,
   With music from his loving hand;
He is not dead—he cannot die!
   A king of men he still returns,
And rules as erst with spirit high,
   The land of ROBERT BURNS.

In clouds of glory dash'd with rain,
   With heavenly light-gleams bound and furled,
From *his* high Caucasus of Pain,
   He casts a song-wreath round the world;
And weakest souls beneath his spell
   Have gathered strength as he who spurns
The might of tyrants—it is well!
   God bless you, ROBERT BURNS!

To Robert Burns.

Dedicated to GEORGE CALL, Esq., *President of the Clachnacuddin Burns Club of New York, an ardent admirer of the Poet.*

JOHN PATTERSON.

ONCE more we meet on this thy natal day,
    Leal-hearted Clachnacuddin lads a few,
    And lassies fair—none Scottish more than they—
To weave a garland round thy name anew,
    And praise that genius grand which from the plough
    Raised thee to stand, their peer, the great among,
The stamp of Fame immortal on thy brow,
Our own great ROBERT BURNS, the King of Scottish song.

Where dwells to-night a son of Scotia dear ;—
    An isle that's nursed at South Pacific's breast
May be his home, where Nature's lavish cheer
    Impels him to partake, and there to rest ;
Or on its grip, grim Arctic's belt may hold
    As in a vice, which, tight'ning, threatens death ;—
Where'er this night his shelter be, thy bold,
Inspiring lays, O BURNS, sound forth with joyful breath.

The carvéd granite shaft, the sculptur'd bust,
    The marble column rare, e'en bronze or brass,
And Art's all stately works which mortal dust
    Immortalizes with, are but as grass

To that grand Alpine range which tops the clouds,
   And hath its roots intrench'd in Scottish hearts,
Of love, perennial love, for thee, which crowds
All other bards to take their place in minor parts.

And as the years roll by, and older grows
   That fame which was begun near bonnie Doon—
Whose banks will bloom though wrung forth there be throes
   From rueful hearts when past in Love-day's noon,
And lovers false with other charms are paired—
   Like wine which richer, rarer grows with age,
So upward goes—one thought by millions shared—
Thy tower of Fame, as plumb'd by Time's unerring gauge.

As written language on the parchment's face
   Bequeaths to few rich gifts from gen'rous friend,
So thou, on thy pathetic heart, canst trace
   Thy title clear to gems of thought, whose trend
The truth, disrobed, with master hand to paint,
   And from the strutting form of pompous pride
Its tinsel trappings rend, the gilded saint
Unmask, and brush the fulsome shams of life aside.

O BURNS, thou hast unlearn'd the menial task
   To cringe and fawn to popinjays of rank,
Or measure men by sordid rule, or bask
   'Neath rays from gold that claims another's bank;
" A man's a man " will future ages sing,
   And generations will its logic see
With clearer eye than ours, and they shall ring
With joy thy praises down to their posterity.

'Tis not alone the sounds of human woe
    That promptly wake thy lyre's impassioned strain ;
Nor yet, alone, the tears that overflow
    The sad, grief-laden eyes of those who drain
The cup " man's inhumanity to man "
    Is potter of, which rouse to grandeur's sphere
Thy just, indignant ire, that brothers can
Their brothers harm the few short years they sojourn here.

The wounded hare, that limps with less'ning speed
    As drops its life's blood on the purple heath ;
The poor evicted mouse, whose home—the meed
    Which toil and anxious care are crested with—
In ruins is, by ploughshare's cruel blow,
    Alike set up mind's cam'ra to thy soul,
Its burning love to show, whose radiant glow
Will e'er a grateful world illume from pole to pole.

" To Mary in Heav'n " some their lasting love
    To pledge beyond the shadow of the bier ;
And some whose love's fruition ranks above
    The wealth this world can give—to-night are here ;
And both do pity him whose Nature's touch
    So dormant be, that joy and grief, by turns,
Thy matchless lays do not evoke, or such
Whose inmost souls unthrill whene'er we mention—BURNS.

The Grave of Burns.

ROBERT NICOLL.

BY a kirkyard-yett I stood, while many enter'd in,
Men bow'd wi' toil and age—wi' haffets auld an' thin ;
An' 'ithers in their prime, wi' a bearin' proud an' hie ;
An' maidens, pure an' bonnie as the daisies o' the lea ;
An' matrons wrinkled auld, wi' lyart heads an' grey ;
An' bairns, like things o'er fair for death to wede away.

I stood beside the yett, while onward still they went.—
The laird frae out his ha', an' the shepherd frae the bent :
It seem'd a type o' men, an' o' the grave's domain;
But these were livin' a', an' could straight come forth again.
An' of the bedral auld, wi' meikle courtesie,
I speer'd what it might mean? an he bade me look an' see.

On the trodden path that led to the house of worshipping,
Or before its open doors, there stood nae livin' thing ;
But awa' amang the tombs, ilk comer quickly pass'd,
An' upon a'e lowly grave ilk seekin' e'e was cast.
There were sabbin' bosoms there, and proud yet soften'd
    eyes,
  An' a whisper breathed around, "There the loved and
    honour'd lies."

There was ne'er a murmur there—the deep-drawn breath
    was hush'd—
And o'er the maiden's cheek the tears o' feelin' gush'd ;
An' the bonnie infant face was lifted as in prayer ;
An' manhood's cheek was flushed wi' the thoughts that
    movin' were :
I stood beside the grave, and I gazed upon the stone,
And the name of "ROBERT BURNS" was engraven there-
    upon.

Ode

Written for and performed at the Celebration of Robert Burns' Birthday, Paisley, 29th Jan., 1807.

ROBERT TANNAHILL.

Recitative.
WHILE Gallia's chief, with cruel conquests vain,
Bids clanging trumpets rend the skies,
The widow's, orphan's and the father's sighs,
Breathe, hissing through the guilty strain ;
Mild Pity hears the harrowing tones,
Mixt with shrieks and dying groans ;
While warm Humanity, afar,
Weeps o'er the ravages of war :
And, shudd'ring, hears Ambition's servile train
Rejoicing o'er their thousands slain.

But when the song to worth is given,
The grateful anthem wings its way to heaven ;
Rings through the mansions of the bright abode,
And melts to ecstasy the list'ning gods ;
      Apollo, on fire,
      Strikes with rapture the lyre,
      And the Muses the summons obey ;
      Joy wings the glad sound
      To the worlds around,
      Till all Nature re-echoes the lay,—
Then, raise the song, ye vocal few,
Give the praise to merit due.

ODE.

Song.

Tho' dark scowling Winter, in dismal array,
  Remarshals his storms on the bleak hoary hill,
With joy we assemble to hail the great day
  That gave *birth* to the *Bard* who ennobles our isle.
Then loud to his merits the song let us raise,
Let each true Caledonian exult in his praise ;
For the glory of genius, its dearest reward
Is the laurel entwin'd by his country's regard.

Let the Muse bring fresh honours his name to adorn,
  Let the voice of glad Melody pride in the theme,
For the Genius of Scotia, in ages unborn,
  Will light up her torch at the blaze of his fame :
When the dark mist of ages lies turbid between,
Still his star of renown through the gloom shall be seen,
And his rich blooming laurels, so dear to the Bard,
Will be cherish'd for aye by his country's regard.

Recitative.

YES, BURNS, "thou dear departed shade !"
When rolling centuries have fled,
Thy name shall still survive the wreck of time,—
Shall rouse the genius of thy native clime ;
Bards, yet unborn, and patriots shall come,
And catch fresh ardour at thy hallow'd tomb—
  There's not a cairn-built cottage on our hills,
    Nor rural hamlet on our fertile plains,
    But echoes to the magic of his strains,
  While every heart with highest transport thrills :
    Our country's melodies shall perish never,
    For BURNS, thy songs shall live for ever.
      Then, once again, ye vocal few,
      Give the song to merit due.

ODE.

Song.

Hail, ye glorious sons of song,
Who wrote to humanize the soul !
To you our highest strains belong,
Your names shall crown our friendly bowl :
    But chiefly BURNS, above the rest,
      We dedicate this night to thee ;
    Engrav'd in every Scotchman's breast,
      Thy name, thy worth, shall ever be !

Fathers of our country's weal,
Sternly virtuous, bold and free !
Ye taught your sons to fight, yet feel
The dictates of humanity :
    But chiefly BURNS, above the rest,
      We dedicate this night to thee ;
    Engrav'd in every Scotchman's breast,
      Thy name, thy worth, shall ever be !

Haughty Gallia threats our coast,
We hear their vaunts with disregard ;
Secure in valour, still we boast
" *The Patriot and the Patriot Bard.*"
    But chiefly BURNS, above the rest,
      We dedicate this night to thee ;
    Engrav'd in every Scotchman's breast,
      Thy name, thy worth, shall ever be !

Yes, Caledonians ! to our country true,
Which Danes or Romans never could subdue ;
Firmly resolv'd our native rights to guard,
Let's toast " *The Patriot and the Patriot Bard.*"

Burns Remembered.

Rev. Arthur J. Lockhart,
The "Bard of Acadie."

Involved in cloudy vapours gray,
The re-appearing king of day,
Now struggling makes his wintry way
    To wake the morn
When blithest bird of clearest lay—
    Our Burns was born.

With snowy winds abroad to rave
Wild Nature piped a frolic stave,
And rough and hearty welcome gave
    Her favourite boy,
Who should misfortune's storms outbrave—
    Its bolts defy.

The babe she clasped in her rude arms
And nursed him with her smiles and storms,
Moved to wild raptures and alarms
    His minstrel soul,
And held him by her frowns and charms
    In her control.

His was her treasury—her time
Of falling leaves and frosty rime ;

The budding season's singing prime
        With sunshine rife ;
And that " true pathos and sublime
        Of human life."

But him she did not shield from woe ;
Teaching his fiery heart to know
Of tears the bitterest overflow :
        Yet hence there came
The keener sense, the " friendly glow
        And softer flame."

Hail to thee ! chief of Scottish bards !
And first among a world's regards !
Thy music, wed to noble words,
        Goes the world o'er ;
Thy pastoral notes, thy deep heart-chords,
        Sweeten each shore.

We love each song to Nature true,
Like dawn and sunset to the view
Familiar, olden—ever new
        And ever sweet,
As the dear daisy in the dew,
        Meek, at thy feet.

From " burn " and " brae " thy coming brings
Thoughts of all bright and joyous things :
The hawthorn blooms, the merle sings
        Aloud, and—hark !
Singing to his blue heaven upsprings
        The morning lark !

Alas! that e'er a harp so fine
That, swept with ardour so divine,
Could make the lowly virtues shine
    Like stars on high,
Should sound at Passion's soilèd shrine
    So witchingly!

But let us not, in canting strain,
Of *man* or *poet* here complain;
Ours in his nobler song the gain
    We gladly share;
If his the error, his the pain—
    Let us beware!

Alas for ill! yet can we soon
Forget the charm of "Bonnie Doon?"
While Afton soughs her gentle tune
    In solitude,
The soul must bless thy cheerful boon—
    Thy melting mood!—

Thy poet-scorn of mean and low,
Of titled fool, and glittering show,
Thy power to feel the social glow
    And fervid flame;
Unknown are they who do not know
    Thy magic name!

O music spirit! "child of air!"
What generous heart but thou art there!
What chord, from rapture to despair,
    But thou didst move!
Yet on thy front dost chiefly wear
    FREEDOM and LOVE!

Robin Burns.

An Anniversary Rhyme.

ROBERT FORD.

ANCE mair an honour'd day is here,—
A day to Scotchmen ever dear;
An' wale o' billies arena sweer,
      To kythe hobnobbin',
An' drink the toast wi' rousin' cheer—
      "Hip! hip! for ROBIN!"

For ROBIN BURNS, sae true's he said,
He was a rantin' rovin' blade,
The pride o' ilka man an' maid,
      Ower Scotland lang.
For mony a rousin' poem he made,
      An' heart-sweet sang.

He tauld the thochts o' dogs an' men,
To stane an' lime ga'e langwidge plain,
The cottar's cosie but an' ben,
      Saw an' admired,
Till Death himsel' he seem'd to ken,
      Our bard inspired.

"Ye banks an' braes o' bonnie Doon,"
He lent thae words a lo'esome soun';

ROBIN BURNS.

RAB had a heart as weel's a croon,
  An' baith were braw.
His love lilts gae the heart-strung stoun'
  O' grit an' sma'.

Oh ! wha can read his "Mary, dear,"
Without a sympathetic tear?
An' wha can nerveless sit an' hear
  Bauld " Scots wha hae ? "
Sic strains a coward's heart wad cheer,
  To face the fae.

An' syne to read the pawky story
O " Willie's wife," the fulsome sorra,
Or follow " Tam," a' in his glory,
  Ream-fu' o' liquor,
We hotch, an' pech, an' lauch, an' roar aye,
  An' screech an' nicher.

Sae gleg's the wit o' ROBIN BURNS,
He gars us lauch an' greet by turns ;
We lauch wi' him, an' when he mourns,
  We mourn wi' him ;
E'en noo I feel his influence stirrin's—
  E'en noo I see him.

He eggs me on to drive my quill aye,
An' prog ilk sneerin' censor billie,
Wha brands him loon, an' strives to sully
  His deathless fame ;—
Oh, critics ! learn to shun the folly
  O' sic a game.

Ilk' Scotchman worthy o' the name,
Outower the sea, or snug at hame
In cot or ha'—a' ranks the same—
    His temper turns
'Gainst ilka wicht wad blight the fame
    O' ROBIN BURNS.

An' lang may Scotland lo'e her bard,
  is memory gi'e its due reward,
For while the daisy decks the sward,
    An' heads have harns,
The matchless sangs will still be heard
    O' ROBIN BURNS.

Burns.

Glasgow Ballad Club Paper, January 25th.

DAVID WINGATE.

"LET me sing a Birthday Ode,"
  Thus does each adorer cry
  When this natal day draws nigh.
Should Apollo deign to nod,
  Straight is raised the frenzied eye—
Spins the humming top of rhyme,
All about this natal time.

But when I my boon besought,
Grave Apollo nodded not,
And, alas! I well could see
That all the Muses stared at me;
And the whispered wonder came
"What is this aspirant's name?"
Down my head in sorrow hung,
And the Ode remains unsung.

Sitting, then, with soul subdued
To a vexed and fretful mood;
Why, I said, an Ode at all,
That may flat as snow-flakes fall?
Why the yearly rite maintain
Of an eulogistic strain,

Setting *one* within the maze
Of an oft-repeated praise,
Round and round to toil, astray,
On a feastful natal day?
While his peers impatient sit,
Longing till electric wit
Forth shall leap with flash divine
'Mong the walnuts and the wine,
Till the Ode, again achieved,
Leaves their tortured souls relieved.

Nay, although it fell not flatly
As the snow-flake falls, but, patly,
Touched the temper of the time,
And in warm and lofty rhyme
Told again the well-known story—
What increase were there of glory
To the Bard whose songs to-night
Shall the gravest cares make light?

So, imagined be the strain
That should tell you o'er again,
How the man we meet to honour
Woo'd the maiden, Fame, and won her.
How the test of trouble proved him!
How he loved, and how men loved him!
How the fervour of his pen
Thrilled, and thrills the souls of men!
How he woke by rill and river
Echoes that will ring for ever;
How whate'er he stooped to name
Shares the marvel of his fame;
How of him it can be said
Nothing that he touched is dead!

How, although detractors cry
" Much that he has touched *should* die,"
Yet where'er his foot has been
Plodding pilgrims still are seen,
Who believe there yet remains
A gleaning of ungathered grains,
Verses bold, and warbling sweet,
Which will make his fame complete.

Thus, in an imagined strain,
Each may tell the tale again
In the pause of Speech and Song;
Or, fancy-led, may roam among
Hallowed scenes by Ballochmyle,
Logan Water, Doon, and Kyle;
The banks of Ayr or Afton Braes;
And where'er the fancy strays;
Singing bird or blossomed sod,
Be the subject of an Ode;
Each one musing as he may
On the Bard we praise to-day.

On Anniversary of Birth-Day of Robert Burns.

(1844)

SARAH PARKER DOUGLAS.—("*The Irish Girl.*")

ALL hail the day so sacred in hamlet, hall, and isle—
The day that ushered into birth the matchless Bard of Kyle !
All hail the day we celebrate throughout the Poet's land,
Whose name's a dear familiar word in every household band !
All hail the day, from other days so proudly set apart
In honour of the Bard of bards, he of the great warm heart—
A heart upon whose altar burned the Muse's sacred fire,
From whence sprang, touched with living coal, each number of his lyre !

Inspired, he gleaned from Nature's page, what Nature's author spread,
And gave to words of melody the glowing truths he read.
The sparkling dew, the balmy breeze, still eve, and blushing dawn,
The rippling stream, the waving corn, and daisy on the lawn :

Drew from his gushing soul those lays which charm and
    melt and thrill ;
And sweet as breath of Summer morn their incense meet
    us still.
Oh ! cheering, soothing, grateful strains, with Truth's
    bright pearlets decked,
By you inspired, e'en poortith stands with honest head
    erect.

Of charming, thrilling, truthful tone, was Heaven's great
    gifts divine,
The sacred lyre, the talent bright, destined for aye to shine,
All potent to delight and cheer, wake sorrow, pity, love,
And elevate the human mind each grovelling thought above.
Ah ! who would say the minstrel failed his mission to
    fulfil—
Sought rest inglorious on his lees, or let his harp lie still ?
He laid him with the early dead, for brief his span of life,
Yet stored the world with deathless song whilst battling
    'midst its strife,

In which the love-lorn bosom finds a sympathetic voice ;
The merry, what can laughter rouse, and make the heart
    rejoice ;
The humbly pious, what can bear the raptured soul from
    earth,
While gazing on the week's last scene around the cottar's
    hearth.
But what are words to tell the power his wizard harp
    possessed,
To conjure up, create, make bare, the secrets of the breast ;
And, still, his lays can sink a peer, if heartless, to the dust,
And elevate the man of soul who toils to win a crust.

His nature, far beyond a lie, revered religion true ;
But hypocrites, donned in *her cloak*, he could not bear to view.
To them the satire of his song the small-cord scourge became,
Till smarting and unmasked they stood, exposed to scorn and shame.
He erred—like many of his day—perhaps, more quick confessed,
Repented and atoned with all the fervour of his breast ;
Be't as it may, the glorious traits, the genius we recall,
Proclaim our Bard the noblest, best, take him for all in all !

With swelling hearts and tearful eyes, with blended love and awe,
Assemblies meet to honour him—"the Bard that's noo awa' ! "
And never yet was homage given to more exalted worth
Than his, who left such priceless gems to ling'rers on the earth.
Then, hail the day old Scotia's sons have calendered with pride—
The natal day, which thousands greet in many a land beside,
When ROBERT BURNS—the proudest name on Scotland's record traced—
Was given—a boon, to light and cheer and beautify life's waste.

Burns on his Death-bed.

W. M'DOWALL.

"A night or two before Burns left Brow, he drank tea with Mrs. Craig, widow of the minister of Ruthwell. His altered appearance excited much silent sympathy; and the evening being beautiful, and the sun shining brightly through the casement, Miss Craig (afterwards Mrs. Henry Duncan) was afraid that the light was too much for him, and rose with the view of letting down the window blinds. Burns immediately guessed what she meant; and, regarding the young lady with a look of great benignity, said—'Thank you, my dear, for your kind attention; but, oh! let him shine! he will not shine lang for me!' The poet died a few days afterwards at Dumfries."—LAND OF BURNS.

SOON will life's weary whirl be done,
    And I shall reach the peaceful grave;
Soon shall my latest sands be run,
    And Passion's tempest cease to rave.
The goal of death and darkness won—
    This chequered scene no more to see;
Yes, let me view again the sun—
    It winna shine sae lang for me!

Unbind the veil that hides his face,
    And draw yon envious screen aside,
Then shall his gladsome radiance chase
    The mist which owre my couch preside,

BURNS ON HIS DEATH-BED.

Ere yet the ebon gates are barred
    Upon my hours of grief and glee :
Sweet sun, my earnest cry regard—
    Ye winna shine sae lang for me !

Break forth as thou were wont to shine
    When in thy glorious light I trod
To trace the links of love divine
    From nature up to nature's God.
Meet emblem of that Mighty One,
    Thy face reveal, and seem to be
A token of His mercy shown,
    For nane can need it mair than me !

Nae mair I see thee paint the plain,
    Or pierce the leaf-embattled shade,
Or mirrored in the trembling main,
    Or glistening in each dewy blade.
But thou canst make the clay-built cot
    Seem blythsome as yon lily lea ;
Not all forlorn the Poet's lot
    Since thou dost shine aince mair on me !

Thou stay of life, and source of light !
    How doubly dear thy presence now,
When shadows, as of endless night,
    Are gathering o'er my throbbing brow !
Prized wert thou in my songful prime—
    And precious must thou ever be ;
Tho' swiftly comes a mirk, mirk time,
    When thou shalt shine nae mair for me !

The sunless grave ! no straggling ray
    Of thine can reach its dread recess ;

Nor would the soul-deserted clay
  Be conscious of its warm caress.
Yet grieve I not, by care opprest,
  To meet the doom I soon maun dree ;
Since, though it shade thy beams sae blest,
  'Twill scatter far the clouds frae me !

Robert Burns.

Seven Sonnets written for the Poet's Birthday Festival.

MALCOLM TAYLOR, JR.

I.

A POET was a prophet deemed of old—
   The singer then was noted as the seer,
   And dared to pierce with soul perceptions clear,
The Future's vail, to have its scenes foretold;
So I, like priveleged, would now make bold
   To draw the curtain Past, each fold a year,
   That time with vandal touch has mildewed sere,
Until a century has back unrolled—
   And lo ! what scene bursts on my spirit sight—
An humble cot of clay, with roof straw-thatched,
Whose lowly entrance swinging wide, unlatched,
   Reveals the event we celebrate to-night—
There, on a cubby bed, one winter's morn,
The infant ROBERT BURNS was happily born.

II.

Now let me, with my pen's weird wand, forsooth,
   Waive by the windings of his young life path,
   The petty trials he had, as each child hath,
Till soon we see him as a reaper youth :
When, bending low beside some winsome Ruth,

To bind with wheaten gyves the levelled swath,
Or gathering up the golden aftermath,
He tried to sing the love he felt in truth ;
Then woke the poet's spirit in his form,
Moved was his hand to touch the latent chords
That longed to give expression fair in words
　To what his heart felt in affection warm ;
And as he told his love in lilted line,
He wooed the willing Coila, muse divine.

　　　　　　　　III.
Next to my retrospect is he revealed
　The farmer poet, driving team abreast
　And plough-share deep, while sweetly he exprest
His sentiments on Nature seen afield ;
And thus he tilled the fertile soil, to yield
　Him honours great for merits well possest,
　Alike from palaced wearer of a crest,
And appreciative peasant in his bield.
　"Lines to a Mouse," "Lines to a Mountain Daisy,"
"Poor Maillie's Elegy," served to excite
The stoic's sympathies with pure delight,
　And earned in fair return the lavish praise he
Received as an adept in Poesy's art—
A man of feeling, near to Nature's heart.

　　　　　　　　IV.
Thus from the harvest-field, erst-while unseen,
　Arose our laverock RAB, dun coated, shy,
　Who on the ladder-rounds of Song, full high
Did mount, impulsive, with majestic mien,
Through clouds of circumstance, to sing serene,
　Exultant in the literary sky ;
　Awaking all the people far and nigh,

Who wondered what bird coming on the scene
   So charmed their senses with sweet dulcet-strains,
Till plaudits from the critics glad, elate,
As echoes rose, to wide reverberate,
   And reach unto the end of Earth's domains,
While up he soared to Ambition's dizzy height,
   And bathed his wings in Fame's supernal light.

V.

And now behold him, Fashion's pampered child!
   The Pet of Wealth! The social board around
   His favoured friends did reverence profound,
While he, with his own songs, the time beguiled
Till, with that Circe, Pleasure's draught grown wild,
   Our laverock RAB soon had his sad rebound,
   And, faulty, fell back to the common ground,
To sink from sight, in poverty exiled;
   But though was smirched with shame in touching dross
The form that housed his soul, above mere pelf,
Yet crushed not was the better part of self;
   From human failings suffering no loss,
His songs lived on and lingered, still sublime,
Through all the echoing corridors of Time.

VI.

Yes, like the thrush, he in a sonnet framed,
   That e'n in Winter's dearth yet sang elate—
   A birthday prophesy of his own fate—
His lilted love will rise, when e'er is named
The People's Bard; aye, all whose grandsires claimed
   A drop of Celtic blood will celebrate
   As we do now his natal day in state,
And drown in Lethe's tide what could be blamed.
   As said one time the dame who gave him birth,

Viewing the monument at his grave-head,
"Puir ROB, ye asked the world to gie ye bread
An' they gied ye a stane to show your worth."
But more than granite shaft, the Scottish tongue
Will keep his memory forever sung.

VII.

Thus have I, with a prophet's after-sight,
Retraced anew the life-line of a bard,
Who, from a common tiller of the sward
Peered up, to shine in all his talent's might,
Among the gifted sons of genius bright ;
And now let us forget the faults that marred
His day, which of the flesh, served to retard
His spirit in its far transcendent flight,
While, in good fellowship, we eat and sup
Due homage to his name, remembering
"A man's a Man for a' that," as we sing
His "Auld Lang Syne," and quaff a kindness cup
In memory of RAB, our Bard and Brither—
Since "we are a' John Thamson's bairns" the gither.

To Burns.

David Vedder.

I THOUGHT on thy name, so beloved, so adored,
   'Neath each clime of the earth, sweetest Bard of the North;
Of the heights so sublime where thy spirit had soared,
   And the rapturous strains which thy muse bodied forth.

I thought on the sorrows which chequered thy youth,
   On the early misfortunes with which thou hadst striven;
When thou drank'st from the crystalline fountain of truth,
   And inhaled, unalloyed, inspiration from Heaven.

It was *then* that thy masculine fancy took wing,
   And soared like a bird to the summit of Fame,
Pouring warblings as sweet as the music of spring,
   And pure as the Devon's meandering stream.

It was *then* that thy lyric enchantments were sung,
   And each feeling bosom its sympathy spoke;
Thy harp, like a seraph's, melodiously rung;
   For the hand of a master its music awoke.

The great and the noble—by apathy pressed—
   Were stirred by its soul-thrilling music divine;
Even noble-born beauty its magic confessed,
   And Fashion taught Dulness to bow at thy shrine.

But they left thee to wrestle with want and with woe,
   Thy prospects all blighted, thy feelings all sered :
Like a meteor amongst them awhile thou didst glow,
   Like a meteor, alas ! which too soon disappeared.

   .   .   .   .

I thought on the column his genius had raised,
   On the dark dreary grave where his relics repose,
On his sensitive bosom which sorrow had seized,
   On his progress through life, on his loves and his woes.

I thought on his hours of convivial bliss,
   With the friends of his heart, round the magical bowl ;
On the conjugal rapture—the heart-thrilling kiss,—
   And I wept like a woman in fulness of soul.

Sweet Bard, thy renown shall for ever increase ;
   Whilst genius is prized shall thy merits be sung ;
The star of thy fame shall in brilliancy blaze,
   Till Nature's funereal knell shall be rung !

Our loveliest maidens shall yearly bestrew
   With flowerets the green turf that pillows thy head ;
And the salt tear of sorrow for ever shall flow
   O'er the spot where our mightiest poet is laid.

For the Anniversary of Burns.

David Vedder.

USHERED by storms and tempests drear,
   Again the auspicious day returns ;
A day to Caledonia dear,—
   The birth-day of immortal BURNS.
No more the beauteous matron mourns,
   No more her tresses sweep the earth,
Her Poet's mighty name adorns
   The happy land that gave him birth !

O ! for a portion of that fire,
   That pathos, strength, and energy,
With which the Poet swept his lyre
   While struggling with pale poverty ;
Then should my muse adventurous try
   The dignified, the daring theme,—
A theme immeasurably high,—
   Even Scotland's mighty Minstrel's fame.

But that can ne'er forgotten be ;—
   He bade her Doric numbers chime,
And struck her harp, whose silver chords
   Shall vibrate till the end of time.
The pealing, rapturous notes sublime,
   That rung from his immortal lyre,
Shall ever ring through every clime,
   Till blazes Nature's funeral pyre !

His lyrics glad the Scottish swains,
   Where Ganges rolls with sullen roar;
His nervous, soul-ennobling strains
   Resound on Hudson's icy shore:
Beyond the Andean mountains hoar,
   Where sacred Freedom's banners blaze,
Our countrymen his loss deplore,
   And yearly crown his bust with bays.

His satire was the lightning's flash
   Which purified our moral air;
His war songs were the thunder's crash
   Which stirred the lion in his lair:
He painted Scotland's daughters fair,
   All beauty, tenderness, and light,
Like verdant wreaths of flowerets rare,
   With summer dews hespangled bright.

Then let thy heath empurpled plains
   With Tuscan vales for ever vie,
And, Scotland, may thy dulcet strains
   Still rival Tuscan melody:
Let thy maternal tears be dry,
   For though his radiant course be run,
The astonished world with plaudits high
   Proclaims him thine illustrious son!

Freedom's Bard.

John Kelso Kelly.

> But still the burden of his song
> Is love of right, disdain of wrong ;
>   Its master chords
> Are Manhood, Freedom, Brotherhood ;
> Its discords but an interlude
>   Between the words.
>   —Longfellow.

Beside the banks o' bonnie Doon,
    Its flowery braes amang,
There leeved, an' lo'ed, an' wrote, an' wooed,
    The chief o' Scottish sang ;—
A blithe yet melancholy chiel,
    Wi' mair o' brain than gear,
Wha to Parnassus tap did spiel,
    By genius' pathway clear.

Chorus.
    Then strike the harp to Robbie's fame,
        Till ower the warl' is heard
    The music o' that honoured name,
        The name o' Freedom's bard.

What though he was a wilfu' bairn,
    Dame Nature's simple son—
He wrote the lessons we may learn
    O' noble wark begun ;

An' bravely did he spiel life's brae,
  Tho' Fortune him beguiled ;
Like ither men he aft was wae,
  But aye in sang he smiled.

Tho' puir in purse, yet rich in min',
  He mortal face ne'er feared ;
To honest man he aye was kin',
  Nor what his rank e'er spiered.
Sae, independent—Labour's son—
  He scorned to ben' the knee
To wealth or rank or fame alone,—
  For honest pride had he.

Hypocrisy he couldna brook—
  That imp o' Satan's van ;
When he but spak', its black ranks shook
  Afore ae honest man ;
An' sic a peltin' never gat
  Ilk graceless, godless loon ;
He stormed their camp, an' laid it flat,
  An' blew a Gospel soun'.

He sang the Kirk, he sang the State,
  He sang o' man freeborn,
An' wae betide his silly pate
  Wha was o' sense forlorn ;
He sounded Superstition's knell,
  Struck death to Slavery ;
When his indignant soul would swell,
  Farewell to Knavery.

He sang the Brotherhood o' a',
  He sang its comin' day,

An' hailed the glory that would fa'
　Frae its first dawnin' ray;
An' to the latest breath o' life,
　Mid pleasure an' mid pain—
A warrior in a noble strife,
　He focht wi' micht an' main.

But sair forfoch'en, Death ere lang
　Took pity on his wae;
Nae mair was heard the Bardie's sang,
　Nae mair in manly fray
Was seen that buirdly, honest chiel,
　Wha lo'ed an' suffered lang;
Wha to Parnassus tap did spiel,
　An' leeves in Scottish sang.

Chorus.

Then strike the harp to ROBBIE's fame,
　Till ower the warl' is heard
The music o' that weel-kenned name,
　The name o' Scotia's Bard.

The Harp of Burns.

ALEXANDER MACLAGAN.

Ay ! long may Scotia's sons revere
Thee ! Harp of BURNS, thou ever dear !
For many a glad soul-stirring strain,
In banquet-hall, on battle plain,
Has from thy chords in triumph sprung,
Since first thou wert divinely strung !
And therefore do we all revere
Thee ! Harp of BURNS, thou ever dear !

Full many a sweet-toned harp we've heard
Well played, I ween, by skilful bard ;
But never harp, nor lute, nor lyre,
For Nature's native force and fire,
For valour, wisdom, wit, and glee,
Were ever match, brave Harp, for thee !
And therefore do we all revere
Thee ! Harp of Burns, thou ever dear !

In Summer's flowery tints we see
Fair types of thy rich harmony !
We hear among our deep'ning woods
The spirit of thy mournful moods ;
And in our tempests, dark and strong,
The terrors of thy warlike song !
And therefore do we all revere
Thee ! Harp of BURNS, thou ever dear !

THE HARP OF BURNS.

Thy Mountain Daisy's hapless fate,
When "crushed beneath the furrow's weight,"
Thy cowerin' Mousie's " wee bit nibble,"
In ruin'd beil', "o' leaves and stibble,"
Call forth the saddest sigh and start
That ever broke from human heart!
And therefore do we all revere
Thee! Harp of BURNS, thou ever dear!

But when thy numbers, like the blast
Of winter, sweep on fierce and fast,
Again in each wild note we hear
The gathering shout! the charging cheer!
Made England's hosts in terror turn
From victory! Bruce! and Bannockburn!
And therefore do we all revere
Thee! Harp of BURNS, thou ever dear!

Thou liftest Merit's sinking heart,
Thou tell'st him how to bear his part;
Thou prov'st that Honour, Truth, and Right,
Are able yet to cope with Might;
Thou mak'st the tyrant turn in shame
From heavenly freedom's sacred flame:
And therefore do we all revere
Thee! Harp of BURNS, thou ever dear!

Hail! Harp of BURNS! Harp of the North!
Though he is fled who would call forth
The spirit of thy brighter days—
There are who yet will hymn thy praise—
Will of thy matchless glory tell
With glowing hearts—and guard thee well,
With souls that shall for aye revere
Thee! Harp of BURNS, thou ever dear!

Robert Burns: a Centenary Ode.

Rev. William Buchanan, B.A.

We hail to-day his glorious birth, one hundred years ago,
Who taught his brothers o'er the earth to think, to feel, to glow;
Whose independent spirit fires in countless thousands now,
Ay, and will burn till Truth expires—that Roman of the plough !

Who spurned the falsehood of pretence, the insolence of pride,
Who measured men by worth and sense, and not by mere outside ;
Who from the mob that worship state, turned to the sterling few
That honour—what alone is great—the Good, the Just, the True !

Thy story, Burns, a tale unfolds, as thrilling as thy song ;
Oh ! that the age which now beholds might hate thy crying wrong—
The cold neglect, contemptuous airs, the cruel, callous sneers
Proud Dullness towards Genius bears; and worse, perhaps, the tears—

The maudlin tears which only fall as soon as men are dead,
And flow full-coursing down the pall of Bards who wanted
    bread ;
The hypocritic tears accurst, so like their ways and doom,
Who used to kill the prophets first, and garnished next
    their tomb !

He gave a voice to every wood, a tongue to every scene ;
His scorn fell like a lashing flood, electric wit between ;
And satire's blast, rough, roaring, loud, came on like
    driving hail ;
How shrunk the shriv'ring liars, cow'd, behind their rotten
    pale !

His genius like the sun forth shone, to bless our human sight,
And clasp the world in one broad zone of bright and living
    light ;
To banish gloom—alas, that gloom his own career should
    mark !
Yet though the Sun all else illume, the Sun itself is dark.

In BURNS's lustre, oh ! how sweet the wild flow'rs round
    us spread !
The mountain-daisy at our feet lifts up its modest head ;
The broom puts on a yellower flush along our "banks and
    braes ; "
The heather wears a deeper blush as anxious of our praise.

Fairies foot lighter on the lea, and dress in gayer green ;
Fate wears more pleasing mystery when he hold "Hal-
    lowe'en ; "

He waves his wand—witches and ghosts our wizard's spell abide;
He speaks, and lo! the hellish hosts, and "Tam's" immortal ride!

How softly blow those westland winds around the happy spot,
Where married love its dwelling finds, care and the world forgot;
Where peace gives joy a deeper rest and sanctifies our lives,
And each believes his "Jean" the best of women and of wives.

And while that swiftly-footed Time steals on us unaware,
Writes wrinkles on young Beauty's prime, binds Vigour to his chair;
Age looks not crabb'd or forlorn although its strength be gone—
The fresh dew of a second morn is round "John Anderson."

His lyrics stir our British blood wherever Britons toil;
They fell the far Canadian wood, dig the Australian soil;
Where Northern winters hold their reign, and Eastern summers long,
They bind our sons in one strong chain of Sentiment and Song.

Hail Scotia's Bard! Long shall be felt thy lyre so many-stringed;
To soothe, to madden, and to melt, what words like thine are winged?

One age—and do we deem it hard that but one BURNS
 appears?
Nay, men were blessed with such a Bard once in a thousand
 years!

For he shall live, and shall live on, when all those years
 are past;
While harvests wave and rivers run; while pangs and
 passions last;
He'll be till Nature's final hour looks wan in Nature's face,
A name, a presence, and a power to move the human race.

The Auld Brig's Welcome.

Read at the Unveiling of the Ayr Burns Statue.

WALLACE BRUCE.

THE Auld Brig hails wi' hearty cheer—
Uncover, lads, for BURNS is here;
The Bard who links us to all fame,
And blends his own with Scotia's name.

Seven hundred years the winding Ayr
Has glassed my floating image there;
I've seen long centuries glide away,
But ROBIN brought our blithest day.

I heard the Thirteenth's warlike peal
Wake serried ranks of glinting steel;
All wrinkled now, yet in my prime,
I wait with joy the Twentieth's chime.

I cherish weel in memory bright
The glorious deeds of Wallace wight,
And deem the very stones are blest
Which bind the arch his feet have pressed.

I mind the time King Robert's band
With sweeping oar left Arran's strand;
The flame that lit yon beacon hill
All round the world is shining still.

THE AULD BRIG'S WELCOME.

Old Coila's had her share of fame,
Her bead-roll treasures many a name ;
She's had her heroes great and sma',
But ROBIN stands aboon them a'.

The auld clay-biggin' of his birth
Becomes the shrine of all the earth ;
The room where rose the Cottar's prayer—
The proudest heritage of Ayr.

No starlit sky, no summer noon
But kens the banks o' Bonnie Doon ;
No human heart but fondly turns
Responsive to the Land of BURNS.

Ah, BURNS ! who dares to call thee poor—
Each skylark nest on yonder moor,
Each daisy-bloom on flowery mead,
The lambs that on the meadows feed—

Each field and brae by burn or stream
Where wandering lovers come to dream
Are all thine own. As vassals all
We gather here from princely hall—

From lowly cot, from hills afar,
From southern clime, from western star,
To bring our love—all hearts are thine
By title time can never tyne.

The crowning meed of praise belongs
To him who makes a people's songs,
Who strikes one note—the common good,
One chord—a wider brotherhood ;

THE AULD BRIG'S WELCOME.

Who drops a word of cheer to bless
His fellow-mortal in distress,
And lightens on life's dusty road
Some weary traveller of his load.

Who finds the Mousie's trembling heart
Of God's great universe a part :
And in the Daisy's crimson tips
Discerns a soul with human lips.

We little dreamed when Mailie died
Those tender words would speed so wide ;
Men smiled and wept and went their way,
The prince was clad in hodden gray.

Though but a Brig, it garred me greet
To hear him pour his Vision sweet,
And in one crowning climax seal
His pity even for the Deil.

To see the couthie Twa Dogs there,
Their joys and griefs wi' ither sbare—
A cantie tale, it made me smile
That sic a lad was born in Kyle ;

Who caught the witches in a dance
And bound them all in lasting trance ;
The very land is bright and gay
Since Tam o' Shanter rode this way.

The Auld Brig kens the story well
These rippling wavelets love to tell :
" Ayr, gurgling, kiss'd his pebbled shore "—
A fonder kiss his waters bore.

That raptured hour, that sacred vow
Are love's eternal treasures now ;
Montgomery's towers may fall away,
But Highland Mary lives for aye.

And sweeter still the swelling song
Of loyal love repairing wrong ;
Like mavis notes that gently fa' :—
" Of a' the airts the wind can blaw."

Brave, bonnie Jean ! We love to tell
The story from thy lips that fell ;
The lengthened life which Heaven gave
Casts radiant twilight on his grave.

A noble woman, strong to shield ;
Her tender heart his trusty bield ;
The critic from her doorway turns
With faith renewed and love for BURNS.

She knew as no one else could know
The heavy burden of his woe ;
The carking care, the wasting pain—
Each welded link of Misery's chain.

She saw his early sky o'ercast
And gloomy shadows gathering fast ;
His soul by bitter sorrow torn,
And knew that "Man was made to mourn."

She heard him by the sounding shore
Which speaks his name for evermore,
And felt the anguish of his prayer ;
"Farewell, the bonnie banks of Ayr."

THE AULD BRIG'S WELCOME.

Oh, ROBERT BURNS ! by tempest tossed,
Storm-swept, by cruel whirlwinds crossed ;
Thy prayers, like David's psalms of old,
Make all our plaints and waillings cold.

In weakness sown, yet raised in might,
He wept that we might know the right ;
His sweetest pleasures pain-imbued,
His songs a drama's interlude.

And who dare thrust his idle word
Where God's own equities are heard ?
" Who made the heart, 'tis He alone "—
Let him that's guiltless cast the stone.

We know but this : his living song
Protects the weak and tramples wrong :
Refracting radiance of delight
His prismed genius, clear and bright,

Illumes all Scotland far and wide,
And Caledonia throbs with pride
To hear her grand old Doric swell
From Highland crag to Lowland dell ;

To find, where'er her children stray,
Her " Auld Lang Syne," her " Scots wha hae,"
And words of hope which proudly span
The centuries vast—" A man's a man."

Then welcome, BURNS, from shore to shore ;
All hail, our ROBIN, evermore ;
Though late we greet the Ploughman's name
Full in the morning of his fame.

Dinna Forget.

A Reminder on Burns's Birthday Anniversary, Jan. 25th.

HUNTER MACCULLOCH.

I.

FORGET that time has moved the world away
Six generations from auld Scotia's day,
Whereon she sang by mouth of Minstrel BURNS
Sweet songs and true, to which the heart still turns.
Forget the miracles that man has wrought,
The incarnations of immortal thought :
The steam-winged village o'er the railway whirled ;
The electric voice that clicks across the world :
The magic trumpet that o'erreaches space,
Brings voice to voice, when face is far from face.
Forget the wonders that the school child learns,
To better hear the singing preacher, BURNS.

II.

O, gifted soul ! to Scottish hearts how dear !
Whose stirring strains sound earnest and sincere ;
Who now strikes up the rant, and now the Psalm ;
Now sobs with Mary, and roars out in Tam :
Whose amber wit surrounds the homeless Mouse,
And gives to it an everlasting house ;
Whose humble Cottar, with his simple heart,
Now sits exalted in a niche apart :

Who caught the Jolly Beggars in the act,
And made silk purse of that sow's ear of fact ;
Whose songs were words and music at their birth,
And voice our glory, sorrow, love, and mirth.
O, sterling soul ! whose living words inspire ;
Too great to play buffoon for lord or squire ;
Who cared no more for New Light than for Old ;
Who in the cause of Truth was rash, but bold ;
Whose faith embraced the brotherhood of man ;
Who lived and died a true republican.

III.

Dinna forget, though BURNS is made a text
On which the elect of this world and the next—
The rich and righteous—now delight to dwell,
They come unbidden to the poet's well.
Puir folks alone are BURNS's rightful heirs !
For them he sings, his heart and soul are theirs ;
Their customs, habits, manners, loves, hopes, joys,
The warp and woof his master hand employs.
Dinna forget, for all that folks now say,
When BURNS, the bard, was living out his day,
The guinea stamp did not make current gold
Of precious ingots from his mind's rare mould.
Save for a nine-days' masquerade of power,
The freak, the fad, the fancy of the hour ;
An unco for the Caledonian Hunt—
Of rough adversity he bore the brunt.
They entertained no angel in his case,
But oped the door to shut it in his face !
Dinna forget, were BURNS this day alive,
At his crack trade of critic he would thrive :
From Dr. Hornbooks their pretensions strip ;
The Holy Willies scourge with satire's whip ;

The wealthy "dunderpates" would finely scorn
And learn anew that " man was made to mourn."
Dinna forget, were BURNS alive this day,
With these same bitter things to sing and say,
He still would hear the unco-guid's reproof,
He still would see the gentry stand aloof ;
And, blown about by pride and passion's breath,
Would reach his heart's great longing—after death !
Dinna forget that BURNS could not escape
The fate that follows us in many a shape ;
That which he was he was in sheer despite
Of all our systems' rules of wrong and right.
Dinna forget, no man can master fate,
Howe'er so wise or witty, learned or great,
And Scotia's bard was human to the core ;
He lived and died as BURNS—no less, no more.

IV.

The Scot to whom the world sends greeting,
The Bard we weary not repeating,
The BURNS whose star is fixed, unfleeting,
   In heaven set ;
The man with heart for puir folk beating—
   Dinna forget !

Lines.

Read at a Meeting in Philadelphia on the 61st Anniversary of the Birth of Robert Burns.

REV. JOHN BURTT.

SWEET the Bard, and sweet his strain,
Breathed where mirth and friendship reign,
O'er ilk woodland, hill, and plain,
  And loch of Caledonia ;
Sweet the rural scenes he drew,
And the fairy tints he threw
O'er the page, to Nature true,
  And dear to Caledonia :
But the strain so loved, is o'er ;
And the Bard so loved—no more
Shall his magic stanzas pour
  To love and Caledonia !

Ayr and Doon may row their floods,
Birds may warble in the woods,
Dews may gem the op'nin' buds,
  And daisies bloom "fu' bonnie, O,"
Laddies blythe, and lasses fain,
Still may love—but ne'er again
Will they wake the gifted strain
  Of BURNS in Caledonia !
While, his native vales among,
Love is felt, and beauty sung,
Hearts will beat, and harps be strung
  For BURNS and Caledonia !

Ode

For the Burns Anniversary.

WILLIAM THOMSON.

Oh humble harp ! over whose fitful strings
  My youthful fingers oft have idly roved,
  Help me to sing a name I long have loved—
A name to which my heart in worship clings,
  Whose natal morning now returns—
           BURNS.

HE came to us when Scotland's bards
  Had lost their manly tone,
When Scotland's nobles sought rewards
  For flattery of a throne ;
And raised them to a purpose high
In songs the world will ne'er let die.

In lyrics rare he sang the praise
  Of his loved native land,
Which brightened 'neath his rustic lays
  As 'neath a wizard's hand ;
And set aglow the youthful fancy
By his heart-charming necromancy !

What truth and tenderness combine,
What power and pathos in each line !
What varied subjects claim his dreams—
The banks and braes, the flowing streams,

ODE.

  The little mouse, the piping thrush,
  The daisy 'neath the ploughshare's crush,
  The love of "brither bard an' frien',"
  The love of Mary and bonnie Jean,
  The scene in cottage home at night
  That sets the lamp of love alight!
  His heart was love—his strains reveal
  He had no hate even for "The De'il!"

  His soulful songs decay forbid,
   His fame shall ever stand
  Like an eternal pyramid
   Among life's shifting sand!
  His mingled pathos, wit, and fire,
  All coming ages shall admire!

  Not in the little land alone
   That gave the poet birth,
  His songs are sung, his name is known
   O'er all the sea-girt earth—
  Across the broad Atlantic's wave—
  In lands Pacific's waters lave.

  And from these distant climes
  Men who have loved his rhymes
Have to that little green churchyard with reverent footsteps
  come,
   And with low-bending head,
   In loving sorrow shed
A tributary tear above poor BURNS's tomb.

  Since first he saw the light
  Long years have ta'en their flight,
  And wrong has striven with right,
And battles have been fought and lost and won;

And earth has less of night,
And more of sun ;
But the bright laurel green
Around his brow
Is brighter now
Than it in all the years gone bye has been !

Come, then, all loyal-hearted Scots !
"From Maidenkirk to John o' Groats,"
On this our poet's natal day, and worship at the shrine,
Sing loud his never-dying lays,
And weave of everlasting bays
A newer wreath around his noble temple to entwine !
And sing his name,
And deathless fame,
When the "Januar' win's" are sighing.
The bard is dead—
His soul has fled—
But his song is never dying !

While breezes soft the sweet bluebell shall woo—
While on our moors upstarts the sturdy thistle—
While at the gates of heaven the lavrocks whistle,
While woman trust to man, and man is true—
While o'er the " banks and braes o' bonnie Doon "
The rich-songed mavis darts—
While heather scents the smiling summer noon—
Will BURNS live in our hearts !

And ever as his natal morn returns
Our hearts will tribute pay to glorious BURNS.

Oh humble harp ! over whose fitful strings
My youthful fingers oft have idly roved,
When singing of the bard I long have loved
Pleasure unbounded to my heart it brings !

Coila.

To the Memory of Burns.
Also read at a Meeting held in commemoration of the Poet's birth.

FRANCIS BENNOCH.

AGAIN,—again assembled here,
Are men with hearts and soul sincere,
And eyes whose lustre with a tear
    Is dimm'd for him of Coila;
Yes ! here again a chosen few
Shall pay the grateful homage due
To him, the gentle, kind, and true,
    Who sweetly sung in Coila !

A titled sumph was not his lot:
His birth-place was the peasant's cot:
When men have dotard kings forgot,
    They'll think of him and Coila.
He knew no rules, nor studied art,
By which his lays might reach the heart;
Of Genius' self he seem'd a part—
    A genius born in Coila !

And as for laws, he needed none,
His thoughts were laws,—ay, every one,—
And universal as the sun
    That softly beams in Coila;

And nations yet unborn shall raise
Their voices tun'd to notes of praise
Of him whose thoughts were turn'd always
   On Liberty and Coila !

But where are they who made him wink
At things that were, nor *dare to think*,
But work, and thus coercive sink
   The mighty mind of Coila ?
Yes ! where are they ?—You may ask where.—
In deepest gloom or wheresoe'er
Base souls are found, you'll find them there,
   But not with him of Coila !

If ye his fame in truth would see,
Go search among the brave—the free :
There mark the freest soul,—'tis he
   Who loves the bard of Coila !
When hearts enslav'd for freedom glow,
They feel their wrongs, and strike the blow,
Are *free !*—Go, learn who taught them so ;
   They'll shout,—a voice from Coila !

The wand'ring exile doom'd to roam
O'er deserts wild,—o'er Ocean's foam,—
Far,—far from friendship, love and home,
   Is still consoled by Coila !
He thinks not on the arid plains,
Nor fever raging in his veins,
For crooning o'er old Scotia's strains
   He deems himself in Coila !

When virgin bosoms pant with love,
And dream of bliss, whene'er they rove

By winding stream or balmy grove,
   O ! then they think of Coila.
With passion quivering through the brain
They strive to speak, but ah ! how vain,
'Till BURNS with magic rends the chain—
For who could willing love explain,
   Like him, the bard of Coila ?

Whene'er the social few unite
To spend in joy the festive night,
The wit and wisdom dazzling bright
   Are borrow'd beams from Coila !
Whate'er his theme, the Poet shone ;
The lyre he struck was all his own ;
'Tis broken now,—the Bard is gone,—
   And Genius weeps o'er Coila !

On the Death of Burns.

RICHARD GALL.

THERE'S a waefu' news in yon town,
    As e'er the warld heard ava ;
There's a dolefu' news in yon town,
    For RABBIE'S gane an' left them a'.

How blythe it was to see his face
    Come keeking by the hallan wa' !
He ne'er was sweirt to say the grace,
    But now he's gane and left them a'.

He was the lad wha made them glad,
    Whanever he the reed did blaw.
The lasses there may drap a tear,
    Their funny friend is now awa.

Nae daffin now in yon town ;
    The browster-wife gets leave to draw
An' drink hersel' in yon town,
    Sin' ROBBIE gaed and left them a'.

The lawin's canny counted now,
    The bell that tinkled ne'er will draw,
The king will never get his due,
    Sin' ROBBIE gaed and left them a'.

DEATH OF BURNS.

The squads o' chiels that lo'ed a splore
  On winter e'enings, never ca';
Their blythesome moments a' are o'er
  Sin' ROBBIE'S gane an' left tbem a'.

Frae a' the een in yon town
  I see the tears o' sorrow fa',
An' weel they may in yon town,
  Nae canty sang they hear ava'.

Their e'ening sky begins to lour,
  The murky clouds thegither draw;
'Twas but a blink afore a shower,
  Ere RABBIE gaed and left them a'.

The landwart hizzy winna speak;
  Ye'll see her sitting like a craw
Among the reek, while rattons squeak—
  Her dawtit bard is now awa'.

But could I lay my hand upon
  His whistle, keenly wad I blaw,
An' screw about the auld drone,
  An' lilt a lightsome spring or twa.

If it were sweetest aye whan wat,
  Then wad I ripe my pouch an' draw,
An' steep it weel amang the maut
  As lang's I'd saxpence at my ca.

For warld's gear I dinna care,
  My stock o' that is unco sma'.
Come, friend, we'll pree the barley-bree
  To his braid fame that's now awa,

On Burns Anniversary.

Hew Ainslie.

We meet not here to honour one
   To gear or grandeur born,
Nor one whose bloodiness of soul
   Hath crowns and kingdoms torn.

No, tho' he'd honours higher far
   Than lordly things have known,
His titles spring not from a prince,
   His honour from a throne.

Nor needs the bard of Coila arts
   His honour to prolong;
No flattery to gild his fame;
   No record but his song.

O! while old Scotia hath sons
   Can feel his social mirth,
So long shall worth and honesty
   Have brothers upon earth.

So long as lovers, with his song,
   Can spurn as shining dust,
So long hath faithful woman's breast
   A bosom she may trust.

And while his independent strain
  Can make one spirit glow,
So long shall Freedom have a friend,
  And Tyranny a foe !

Here's to the social, honest man,
  Auld Scotland's boast and pride !
And here's to Freedom's worshippers
  Of every tongue and tribe.

And here's to them, this night, that meet
  Out o'er the social bowl,
To raise to Coila's darling son
  A monument of soul.

What heart hath ever matched his flame ?
  What spirit matched his fire ?
Peace to the Prince of Scottish song,
  Lord of the bosom's lyre !

Lines Written for Burns' Anniversary.

JOHN MITCHELL.

AN' me that beuk awee, guidwife: I think it's ROBIN BURNS,
ıs' lines excite within our breasts such mirth and grief by turns;
while we're sitting by the fire, I'll read a page or twa,
t will to bed-time banish sleep, an' wear the night awa'.

I read you how John Anderson's auld wife his worth would praise,
ıow the Twa Dogs talk'd o' men, and weel they ken't their ways,
ıow Death in his rage wad rail 'gainst Hornbook an' his crew,
doing to our race the thing that he himsel' should do.

: will I read you what Bruce sang when England's gather'd might
ear'd on Bannockburn's proud plain, array'd in armour bright,
:wine the thistle with the rose, without the thistle's leave,
which, as every Scotsman kens, England had cause to grieve.

"Or will I read that darker page how hapless man maun dree
The ills that wait on hoary eild when join'd to poverty;
It aft has tears brought frae thine e'e, when, o'er the words that burn,
Thou lean'd to hear, my guid auld wife, how Man was made to Mourn.

"Just ope the volume where ye please," the gude auld wife replied,
"There's no a page atween its boards but ye hae aften tried,
And I hae listened wi' delight to ROBIN's gleefu' tales,
Wha's lines raise gladness in our hearts, and Nature's face unveils.

"The Mouse! wha cared about a mouse till ROBIN'S mouse appear'd,
And saw its wee bit housie wreck't that wi' sic pains it rear'd;
And, O! in what a thrilling strain has ROBIN sung its waes,
When frae its warm beil forced to rin, an' skulk amang its faes.

"Or Tam O'Shanter, O! gudeman, I've heard it ten times owre,
And aften fancy to mysel' poor Tammie's wilyart glower,
When gazing on the "towzie tyke" wha play'd the pipes sae weel,
That supple Nancy scorn'd to rest, but join'd in every reel.

"The Mountain Daisy? yes, try it, I ne'er heard ought sae fine,
There's beauty in the verra words, there's truth in every line;
And ever since I heard it read I ne'er the wee things see
But I hae min' o BURNS, an' they are dearer far to me.

"Some soulless sumphs may cock their snouts at what our
    bard has said,
But ere his words are lost, our vales, our verra hills will
    fade ;
His lines live in each Scotsman's heart, are woven in his
    tongue,
And generations yet unborn will see his fame still young.

"Thae sumphs may think they're doing right, but wiser
    folk ken weel
They ken nae mair o' Nature's warks than my auld spinning
    wheel,
Or wad they dare to slight the beuk that to the mind
    imparts
The charms that elevate the soul, an' captivate our hearts.

"Then ope the volume where ye please, ye canna gang far
    wrang,
Tho' ye should read Glencairn's lament or some bit canty
    sang ;
We've read the hale o' them before, we'll read them yet by
    turns,
For naething comes amiss, ye ken, that comes frae ROBIN
    BURNS."

So spake the gude auld wife ; and we wha worship at his
    shrine,
Will sing wi' joy the strains that make his name almost
    divine ;
Then loudly toast his deathless name ! it dark oblivion
    spurns,
Till echo 'mid her rocks forget to echo aught but BURNS !

Robert Burns: A Centenary Song.

Gerald Massey.

A HUNDRED years ago this morn,
   He came to walk our human way;
And we would change the crown of thorn
   For healing leaves to-day.

But we can only hang our wreath
   Upon the cold white marble's brow!
Tho' loud we speak, or low we breathe,
   We cannot reach him now.

He loved us all! He loved so much!
   His heart of love the world could hold;
And now the whole wide world with such
   A love would round him fold.

'Tis long and late before it wakes,
   So kindly, yet a true world still;
It hath a heart so large, it takes
   A century to fill.

Ay, tell the wondrous tale to-day,
   While songs are sung, and warm words said;
Tell how he wore the hodden grey,
   And won sweat-sweetened bread.

With wintry welcome at the door
  Did Nature greet him to his lot ;
Our royal Minstrel of the Poor,
  Hid in an old clay cot.

There, in the bonny bairntime dawn,
  He nestled at his mother's knee,
With such a face as might have drawn
  The angels down to see

That rosy innocent at prayer,
  So pure and ready for the hand
Of her who is guardian angel where
  Babes sleep in Silent Land.

There, young Love slyly came, to bring
  Rare balms that will bewitch the blood,
To dance while happy spirits sing,
  With life in hey-day flood.

And there she found her darling child,
  The robust Muse of sun-browned health,
Who nurs'd him up into the wild
  Young heir of all her wealth ;

And there she rock'd his infant thought,
  Asleep with visions glorious
That hallow now the poor man's cot
  For evermore to us :

Disguised angelic playmates were
  Those still ideal dreams of youth,
That drew it on to greatness, there
  We find them shaped in truth !

A CENTENARY SONG.

And there he learned the touch that thrills
  Right to the natural heart of things ;
Struck rootage down to where life heals
  At the eternal springs :

Before the lords of earth there stood
  A man by Nature born and bred,
To show us on what simple food
  A hero may be fed.

No gifts of gold for him, no crown
  Of fortune ready for his brow ;
But wrestling strength to earn his own ;
  It shines in glory now !

He rose up in a dawn of light
  That burst upon the olden day ;
Many weird voices of the night
  In his music passed away.

He caught them, Witch and Warlock ! ere
  They vanish'd ; all the revelry
Of wizard wonder we must wear
  The mask of sleep to see.

Droll humours came for him to paint
  Their pictures ! straight his merry eye
Had taken them, so queer and quaint,
  We laugh until we cry.

Wild music on lone shingly shores ;
  Wild winds that break in seas of sound ;
Sad twilights eerie on the moors ;
  The murdered martyr's mound ;

Dark awful shadows trailing like
   The great skirts of the hurrying storm ;
Wan purple thunder-lights that strike
   The woodlands wet and warm ;

Meek glimpses of peculiar grace
   Where beauty lieth in undress
Asleep in secret hiding place,
   Out in the wilderness ;

Those sunsets where, thro' God's good-night
   To our fair world is smiled, and felt ;
All, all enrich his ear and sight—
   Thro' all his being melt.

He knew the sorrows of poor folk,
   He felt for all their patient pain ;
And from his clouded soul he shook
   A music soft as rain.

For them his eyes would brim with balm,
   Dark eyes, and flashing as the levin—
Grew at a touch as sweet and calm
   As are the eyes of heaven.

So rich in sadness is his breast
   That tenderness, heaven-mirroring, fills,
As lies the soft blue lake at rest
   Among the rugged hills.

And quick as mother's milk will rise,
   At thrill of her babe's touch, and strong
It heaves his heart, and floods his eyes,
   And overflows his song ;

A CENTENARY SONG.

In Life's low ways, and starless night,
  The Poor so often have to creep
Where Manhood may not walk full height,
  And this made ROBIN weep.

But none dare sneer, who see the tear
  In ROBIN BURNS's honest eye,
With all the weakness, it comes clear
  From where the thunders lie.

Such ardours flash from out that dew,
  And quiver in that pearl of pain,
The Spirit of Lightning thrilling thro'
  Its drop of tempest rain!

———

Of all the Birds the Robin he
  Is darling of the gentle Poor;
His nest is sacred, he goes free
  By window or by door:

His lot is lowly, and his wings
  Are only of the homely brown,
But in the rainy day he sings
  When gayer friends have flown.

And hoarded up for us he brings,
  In that brave breast of bonny red,
A gathered glory of the Springs,
  And Summers long, long fled.

Even so all Birds of Song above,
  To which the poor man smiling turns,
The darling of his listening love
  Is gentle ROBIN BURNS:

His summer soul our winter warms,
  He makes a glory in our gloom ;
His nest is safe from all the storms,
  For ever in our home.

Yes, there is such a human glow
  Of life and love in ROBIN's breast,
Its warmth can melt the winter snow
  In Poverty's cold nest.

Auld Scotland's music wandered long,
  And wailed and wailed about the land,
Divinely yearning in her wrong,
  And sorrowfully grand ;

And many touch'd responsive chords,
  But could not tell what she would say ;
Till ROBIN wed her with his words,
  And they were One for aye.

His Ministers of Music win
  Their way where night is all so mirk,
You scarce can see the Devil in
  That darkness at his work,

Or feel the face of friends from foes ;
  But these Song-Spirits softly come,
And lo ! a light of heaven glows
  Within the poorest home.

On either side the hearth they glide,
  And take the empty seat of Care,
Immortal Presences that bide
  In blessed beauty there.

They set us singing at our work,
   Or where no ringing voice is found;
Out smiles the music that may lurk
   In thoughts too fine for sound.

They weave some pictured tints that shine
   Luminous in life's cold grey woof;
They make the vine of Patience twine
   About the barest roof.

More sweet his songs, to him who plods
   Shut up in smoky city prison,
Than to the cagéd lark cool sods
   Cut ere the sun be risen.

The soldier feels them as a spring
   Of healing 'mid the Indian sand;
They gush within him, and they bring
   Such news of the old land !

With them the sailor warms his heart,
   Out on the bitter wintry sea;
With them our serfs ennobled start
   I' the knighthood of the Free !

Ah, how some old sweet cradle-song
   The wayward wandering heart still brings
Home ! home again, with ties as strong
   As Love's own leading-strings.

We hug the homestead, and more near
   The fresh and fonder tendrils twine
To make our clasp more close for fear
   Our dear ones we may tine.

A CENTENARY SONG.

When Hesper thro' some shady nook
   Sparkles on lovers face to face,
Where droop'd lids shade a burning look,
   With beauty's shyer grace—

And holy is the hour for love,
   And all so silent comes the night,
Lest even a breath of faërie move
   That poise so feather light—

Where two hearts weigh, to blight or bless,
   Till swarming like a summer hive,
The inner world of happiness
   With music grows alive—

There as life aches so, heart in heart,
   And hand in hand so fondly yearns,
Love shakes his wings, and soars and sings
   Some song of R0BIN BURNS.

Think how those heroes, true till death,
   In Lucknow listened thro' the strife,
And held what seemed their latest breath
   They had to draw in life,

To hear the old Scots' music dear
   Ask, down the battle pauses brief,
As Havelock's men, with fire and cheer,
   Swept in to their relief—

"*Should auld acquaintance be forgot ?*"
   Thro' flaming hell we come ! we come !
To keep that pledge, not given for nought,
   Around the hearth at home ;

A CENTENARY SONG.

"*We'll take a cup o' kindness*" here,
    For Scotland yet, and Auld Lang Syne ;
Ay, tho' that cup be filled with dear
    Heart's blood instead of wine ;

"*And here's a hand, my trusty friend,*"
    And then it seemed the dear old land
Did burst their tomb, the death-shroud rend,
    And clasp them with her hand.

How dearly ROBIN lo'ed the land
    That gave such gallant heroes birth ;
Its wee blue bit of heaven, and
    Its dear green nook of earth.

And dearer is the purple heath,
    The bonny broom of beamless gold ;
And sweeter is the mellow breath
    Of Autumn on the wold !

Where he once look'd with glorious gaze,
    In all our way-side wanderings,
Shy Beauty lifts her veil of haze,
    And smiles in common things.

The Daisy opes its eye at dawn,
    And straight from Nature's heart so true,
The tear of BURNS peeps sparkling ! one
    Immortal drop of dew !

With eyes a thought more tender, we
    Look on all dumb and helpless things ;
In his large love they stand, as He
    Had sheltered them with wings.

A CENTENARY SONG.

Down by the singing burn we greet
   His voice of love and liberty ;
High on the bleak hill-side we meet
   His spirit blythe and free !

And on this land should foe e'er tread,
   He will fight for it at our side,
Flame on our banners overhead,
   In songs of victory ride.

A hundred years ago to-day,
   The great and glorious stranger came ;
Men wondered as he went his way
   A wild and wandering flame.

The fiercer fire of life confined,
   With higher wave 'twill heave and break,
And higher should the mountain mind
   Thrust up its starward peak :

But often is the kindling clay
   With its red lightnings rent and riven,
And Earth holds up a wreck to pray
   For healing hand of Heaven.

Around his soul more sternly warred
   The powers that smite for Wrong and Right,
And thunder-scathed and battle scarred,
   Death bore him from the fight.

But now we recognize in him,
   One of the high and shining race ;
All gone the mortal mists that dim
   The fair immortal face.

A CENTENARY SONG.

The splendour of a thousand Suns
  Is shining ! and the tearful rain
No more with passionate pathos runs,
  And there is no more pain.

The sorrow and suffering, soil and shame
  All gone ! all far away have passed ;
He sitteth in the heavens of Fame
  Quietly crowned at last.

The prowling ghoul hath left his grave,
  Hush'd is the praying Pharisee ;
His frailties fade, his virtues brave
  Shall work immortally.

The spots on this side of our star,
  We saw because it burned so bright ;
But on the other side they are
  All lost in greater light.

Weep, weep exulting tears that he
  The lowly born, the Peasant's son,
Hath wrought for us imperishably ;
  A peerless place hath won !

And such a crown to bind thy brow,
  Thy glorious child hath gained for thee,
Thou grey old nurse of heroes ! thou
  Proud mother, Poverty !

Look up ! and let the big tears be
  Triumphant touch'd with sparks of pride ;
Look up ! in his great glory we
  Are also glorified.

Or weep the tear that Pity wrings
   To think his brightness he should dim ;
Then 'tis the tear of sorrow brings
   Us nearer unto him :

'Tis here we touch his garment, here
   The poorest or the frailest earns
The right to call him kinsman dear,
   Our brother, ROBIN BURNS.

In fires of suffering far more fair
   We forge the precious bond of love ;
Ah, ROBIN, if God hear our prayer
   'Tis all made well above.

And you who comforted His poor
   In this world, have eternal home
With those He comforteth, His poor !
   In all the world to come.

Dear Highland Mary went before
   To plead for you in saintly sooth,
Whom she remembered when you wore
   The purity of youth.

With those high bards who live for aye,
   Your faults and failings all forgiven—
May there be festival to-day,
   And a great joy in Heaven !

The truth afar off found at last ;
   The triumph rung impetuously
Thro' all that Crystal Palace vast
   Of white Eternity.

A CENTENARY SONG.

Ah, ROBIN, could you but return
  Once more, how changed it all would be ;
The heart of this wide world doth yearn
  To take you welcomingly.

Warm eyes would shine at windows ; quick
  Warm hands would greet you at the door,
Where oft they let you pass heart sick,
  So heedlessly of yore !

And they would have you wear the crown
  Who bade you bear the crushing cross ;
Their glorious gain was all unknown,
  Without the bitter loss :

The cup you carried was so filled,
  The pressing crowd so eager round,
Dragged down your lifted arm, and spilled
  Such dear drops on the ground !

How we would comfort your distress,
  Would see you smile as once you smiled,
And hold your hands in silentness,
  Strong man and little child !

Your poor heart heaving like the waves
  Of seas that moan for evermore,
And tried to creep into the caves
  Of Rest, but find no shore.

Poor heart ! come rest thee from the strife ;
  Come, rest thee, rest thee in the calm,
We'd cry ! come bathe thy weary life
  In Love's immortal balm !

A CENTENARY ODE.

We cannot see your face, ROBIN !
   Your flashing lip ! your fearless brow ;
We cannot hear your voice, ROBIN !
   But you are with us now :

Altho' the mortal face is dark
   Behind the veil of spirit-wings,
You draw us up as Heaven the lark
   When its music in him sings.

With tender awe we feel you near,
   You make our lifted faces shine ;
You brim our cup with kindness here,
   For sake of Auld Lang Syne.

We are one at heart as Britain's sons,
   Because you join our clasping hands,
While one electric feeling runs
   Tho' all the English lands.

And near or far where Britons band,
   To-day the leal and true heart turns
More fondly to the fatherland
   For love of ROBIN BURNS.

Robert Burns.

January 25, 1877.

WILLIAM FREELAND.

ROBIN, ROBIN, lo ! 'tis your day,
   The Janwar Day when you were born !
We love you still, as well we may ;
   For still you give us smiles and tears
   And sweetness to keep green the years,
   To strengthen hopes, to conquer fears :
   And now, in spite of spite and scorn,
   You rear your brow into the morn,
     Bold, unforlorn,
       ROBIN, ROBIN.

ROBIN, ROBIN, with happy mind
   We bless the Heavens that sent you here,
When men were deaf and dumb and blind,
   And could not hear dear Nature sing,
   Nor touch Love's tender human string,
   Nor see in each free man a king,
    Till thrilled by you—man without fear :
   Then woke the land, all eye, all ear,
     Tongue, trumpet-clear,
       ROBIN, ROBIN.

Robin, Robin, behold we come
    To laurel you, our Bard-king, there—
Song-silent 'mid the passionate hum :
  Silent ! nay, but your lips of bronze
  Shall start a living soul from stones,
  With Liberty's eternal tones,
    And wake us from our mean despair
    To breathe the patriot's glorious air,
      Stern, glad, and fair,
        Robin, Robin.

Robin, Robin, ring out your voice ;
    Kindle anew the flame of song ;
And make the common world rejoice
  At right made might, at galling chains
  Falling beneath truth's lightning strains,
  At life made rich by honest gains,
    At tyrants lashed with their own thong,
    And nations trampling lie and wrong,
      Manly and strong,
        Robin, Robin.

THE END.

www.ingramcontent.com/pod-product-compliance
Lightning Source LLC
Chambersburg PA
CBHW031902220426
43663CB00006B/734